Inhuman Power

Digital Barricades:
Interventions in Digital Culture and Politics

Series editors:
Professor Jodi Dean, Hobart and William Smith Colleges
Dr Joss Hands, Newcastle University
Professor Tim Jordan, University of Sussex

Also available:

Shooting a Revolution:
Visual Media and Warfare in Syria
Donatella Della Ratta

Cyber-Proletariat:
Global Labour in the Digital Vortex
Nick Dyer-Witheford

The Digital Party:
Political Organisation and Online Democracy
Paolo Gerbaudo

Gadget Consciousness:
Collective Thought, Will and Action in the Age of Social Media
Joss Hands

Information Politics:
Liberation and Exploitation in the Digital Society
Tim Jordan

Sad by Design:
On Platform Nihilism
Geert Lovink

Unreal Objects:
Digital Materialities, Technoscientific Projects and Political Realities
Kate O'Riordan

Inhuman Power

Artificial Intelligence and the Future of Capitalism

Nick Dyer-Witheford, Atle Mikkola Kjøsen
and James Steinhoff

First published 2019 by Pluto Press
345 Archway Road, London N6 5AA

www.plutobooks.com

British Library Cataloguing in Publication Data
A catalogue record for this book is available from the British Library

ISBN 978 0 7453 3861 3 Hardback
ISBN 978 0 7453 3860 6 Paperback
ISBN 978 1 7868 0395 5 PDF eBook
ISBN 978 1 7868 0397 9 Kindle eBook
ISBN 978 1 7868 0396 2 EPUB eBook

Typeset by Stanford DTP Services, Northampton, England

Simultaneously printed in the United Kingdom and United States of America

Contents

Series Preface vi
Acknowledgements vii

Introduction: AI-Capital 1

1 Means of Cognition 30

2 Automating the Social Factory 68

3 Perfect Machines, Inhuman Labour 110

Conclusion: Communist AI 145

Notes 163
Bibliography 168
Index 200

Series Preface

Crisis and conflict open up opportunities for liberation. In the early twenty-first century, these moments are marked by struggles enacted over and across the boundaries of the virtual, the digital, the actual and the real. Digital cultures and politics connect people even as they simultaneously place them under surveillance and allow their lives to be mined for advertising. This series aims to intervene in such cultural and political conjunctures. It features critical explorations of the new terrains and practices of resistance, producing critical and informed explorations of the possibilities for revolt and liberation.

Emerging research on digital cultures and politics investigates the effects of the widespread digitisation of increasing numbers of cultural objects, the new channels of communication swirling around us and the changing means of producing, remixing and distributing digital objects. This research tends to oscillate between agendas of hope, that make remarkable claims for increased participation, and agendas of fear, that assume expanded repression and commodification. To avoid the opposites of hope and fear, the books in this series aggregate around the idea of the barricade. As sources of enclosure as well as defences for liberated space, barricades are erected where struggles are fierce and the stakes are high. They are necessarily partisan divides, different politicizations and deployments of a common surface. In this sense, new media objects, their networked circuits and settings, as well as their material, informational and biological carriers all act as digital barricades.

Jodi Dean, Joss Hands and Tim Jordan

Acknowledgements

The authors thank each other for a collegial and comradely collaboration. We thank the organizers of the 'Das Kapital at 150: Marx's Critique of Political Economy and the Global Crisis Today' conference at Hofstra University in April 2017, where a panel presentation gave us our first opportunity to articulate some of these ideas. We also thank the organizers of the conference 'Log Out! Worker Resistance Within and Against the Platform Economy', held on 6 March 2018 at the University of Toronto, which inspired much of the 'Heptagon of Struggles' section of Chapter 2. We thank David Castle for so immediately, and then patiently, supporting publication of this book, and the staff at Pluto for their work on it.

Nick Dyer-Witheford thanks his dear wife Anne for countenancing and enlivening the writing of another book. Atle Mikkola Kjøsen thanks his lovely partner Siobhan for all her support and companionship throughout the writing process. James Steinhoff thanks Marcello Guarini, Stephen Pender, Jeff Noonan, Deborah Cook, Philip Rose and Chris Tindale for helping him grow a brain.

Introduction: AI-Capital

THE MOST VALUABLE THING

Capitalism is today possessed by the Artificial Intelligence (AI) question. Consider the Vancouver start-up Sanctuary Cognitive Systems Corporation, which aims to develop 'humanoid robots that can move, speak and think for themselves and interact – as intellectual peers – with real people'. Its owner, Geordie Rose, a quantum computing pioneer, concedes to an interviewer that, while building blocks towards this goal are already in common use in multiplying types of 'narrow AI' and specialized robotics applications, none are remotely close to an 'Artificial General Intelligence' (AGI) capable of full human emulation. His company's mission 'to unlock how human intelligence works and to replicate it on a mass scale' therefore 'sounds like a mind-boggling moonshot'. However, Rose is undeterred, for if he succeeds, it will be 'the most valuable thing ever created. What we're talking about is fundamentally altering the basis of capitalism itself' (Silicoff 2018).

It would be easy to dismiss this quest, were Sanctuary not competing against some of the most powerful capitalists in the world, also striving to produce AGI and related technologies: Elon Musk and his non-profit OpenAI; Vicarious FPC, Inc., backed by Samsung and technology billionaires Mark Zuckerberg and Jeff Bezos; and DeepMind Technologies, acquired by Google/Alphabet, whose owners, Sergei Brin and Larry Page, are patrons of the transhumanist Raymond Kurzweil (2005a), the most famous prophet of a 'technological singularity' in which computers attain human-equivalent intelligence. Sanctuary Cognitive Systems Corp may or may not survive (it is clearly cash-strapped and looking for angel investors). But its story of soaring technological ambition, mission-driven digital entrepreneurship and creepy androids 'like underworld creatures from a Hieronymus Bosch painting' (Silicoff 2018) is symptomatic of the AI-fever sweeping the world market, a fever that also manifests in a burgeoning business literature on AI applications, torrents of conflicting predictions about AI's consequences for employment, utopian speculation on the creation of 'Life 3.0' (Tegmark 2017), and fictions ranging from pulp robo-apocalypses (Wilson 2012, 2015) to complex literary explorations on the new techno-existential horizon posited by AI (Mason 2017).

Defining AI is difficult. Nonetheless, Rose is correct that what is often termed 'narrow AI' is already present in the algorithmic processes that now inform much of everyday life. For warehouse workers or military personnel, such AIs may incarnate in the chassis of a robot delivery vehicle or semi-autonomous killer drone. Most AIs, however, act invisibly in the background of activities conducted on smartphones and computers; in search engine results, social media feeds, video games and targeted advertisements; in the acceptance or rejection of applications for bank loans or welfare assistance; in a call centre inquiry or summons to an on-demand cab; or in encounters with police or border guards, scanning their shadowed screens. In these ways, AI has been with us for years.

Once upon a time, people on the left referred to the regimes of the USSR and Eastern Europe as 'actually-existing socialism' (Bahro 1978), indicating an incipient but imperfect realization of hopes for a new social order. We propose an analogous formulation: 'actually-existing AI-capitalism', designating a phase of experimental and uneven adoption of the technologies in which so many hopes are invested. This phase may be protracted far longer than AI enthusiasts anticipate. It may stagnate, stall out and implode (as 'actually-existing socialism' did). But it could also intensify or expand in a transition either to a significantly transformed capitalism, or to a radically different social formation.

This reference to the fate of socialism brings us to the vantage point from which we critique AI-capitalism. Prophets of a technological singularity expect its arrival around 2045 (Kurzweil 2005a). This would put it 201 years after a critical observer of early industrial capitalism penned his own prediction as to its eventual outcome: 'finally … an inhuman power rules over everything' (Marx 1975 [1844]: 366). The young Marx was not writing about artificial intelligence or robots. He was describing the 'alienation' of workers dispossessed by capital of control over what they made, how they made it, their relations with fellow human beings and of their very 'species-being'. The argument we make in this book is that AI should be seen as the culmination of this process, a moment where the market-system assumes a life of its own. AI, we posit, is 'alien power' (Marx 1990: 716) – the power of autonomous capital. We read AI and Marxism through one another: AI through Marx, because Marx's analysis of capitalism is the most comprehensive critical account of the fusion of commodification and technology driving AI forward today; Marx in the light of AI, because AI problematizes human exceptionalism, agency and labour in ways that profoundly challenge Marxist assumptions, and hence requires careful examination by those who share Marx's aspiration for revolution against and beyond capital.

THREE POLEMICS

The argument of this book interweaves three polemic critiques. The first is a critique of AI as an instrument of capital, with all this entails in terms of both the exploitation in and ejection from waged work of human labour, and the concentration of wealth and social power in the hands of the corporate owners of high technology. Depictions of AI as the outcome of a disinterested process of scientific research are naive. Machine intelligence is the product not just of a technological logic, but simultaneously of a social logic, the logic of producing surplus-value. Capitalism is the fusion of these technological and social logics and AI is the most recent manifestation of its chimerical merging of computation with commodification. Jump-started by the digital experiments of the US military-industrial complex, AI emerged and developed within a socio-economic order that rewards those who own the means for automating human labour, accelerating sales, elaborating financial speculation and intensifying military-police control over potential restive populations.

Whether or not AI may be put to different uses – 'reconfigured' (Bernes 2013; Toscano 2014; Steinhoff 2017) to contribute to or create a different social order – is a question we discuss later. What is apparent is that the owners of the great digital corporations regard AI as their technology – and with good reason, for it is they who possess the intellectual property rights, the vast research budgets, the labour-time of AI scientists, the data and the centres that store it, telecommunications networks, and the ties to an enabling state apparatus that are the preconditions for the creation of AI. It is they and their high-ranking managerial cadres who are in a position to implant their goals and priorities within AI software and hardware, 'baking-in' their values – in practice, the one prime directive, to expand surplus-value – to its design.

There may appear to be surprising diversity of opinion about AI amongst corporate leaders, ranging from ecstatic embrace to apocalyptic warning. What is shared, however, is the tacit agreement that it is they who are to dictate the direction of AI, to determine in their high-level conclaves and privileged conversations with government how (not if) it is to be adopted, and with what balance between wild gladiatorial free-enterprise competition and cautionary ethical regulation and policy safety-nets to prevent unwelcome tumults. Whether it is Sergey Brin endorsing the idea of the singularity while consolidating his company's monopolistic powers to direct it, or Elon Musk warning of AI catastrophe while building (or attempting to build) fully automated factories for the production of self-driving vehicles, or Bill Gates offering feeble robot tax plans (and thereby drawing the instantaneous ridicule of peers), the colourful clashes of corporate personalities cover the more sober reality that these great AI moguls are no more, or less, than the personifications of abstract forces of market calculation that drive towards the maximization of profit. They also obscure the massive hubris of the capitalist class that believes it can control the forces it has unleashed. For we did not quite complete our titular quote from the young Marx: 'finally – *and this goes for the capitalists too* – an inhuman power rules over everything (1975 [1844]: 366, emphasis added).

In making this critique of capitalism's encounter with AI we also, however, take issue with leftist theorists who share such concerns, some of whom, like us, quote Marx in their evaluations of AI. So here we quarrel with interlocutors we respect and have learned from, but with whom we differ. There are two specific left perspectives against which we argue, perspectives that we dub 'minimizing' and 'maximizing' views on AI.

The left 'minimalist' position dismisses current discourses on AI as hype and hucksterism. A more moderate version grants them a limited credibility but insists this is not sufficient to seriously change previous analyses of capital and class (Huws 2014; Moody 2018a). For a strong statement of this minimalist position, we can take Astra Taylor's 'The Automation Charade' (2018), an essay whose basic thesis is that 'the rise of the robots has been greatly exaggerated' (like many authors, Taylor sees AI and robotics as pretty much synonymous, an unfortunate gloss we discuss later). Taylor agrees with our point that technological change, and automation in particular, is not a neutral process, but rather wielded from a position of class power. However, her main argument is that it is not just the actuality of automation, but more its possibility, that is weaponized to intimidate workers. She cites the threats of robotized burger-flippers and touch-screen self-service kiosks by fast food corporations trying quell the Fight for 15 minimum wage movement. Some of those threats proved hollow, and those that have been realized, Taylor points out, still leave lots of workers toiling in McDonald's. In light of this, she proposes 'making our idea of automation itself obsolescent. A new term, "fauxtomation", seems far more fitting' (2018). Socialist feminists, she suggests, have, through their close engagement with domestic toil as unwaged work, a special insight into capitalism's ineradicable dependence on human labour, even where that labour is unacknowledged, unrewarded and conducted by women and racialized minorities. She goes on to stress the way the introduction of machines has intensified, rather than eliminated, work, emphasizing the behind-the-scenes dependence of Silicon Valley's digital platforms on the invisible work of figures such as content moderators. Against this background, Taylor takes as a clarion-call moment of insight a response she reports from the famous Marxist feminist theorist, Silvia Federici, to a conference question about capital's tendency to generate 'surplus populations': 'Don't let them make you think that you are disposable.'

Many of Taylor's points are excellent; we expand on some of them later, especially in Chapter 2, where we discuss the labour conditions of AI automation. We concur that 'automation has an ideological function as well as a technological dimension', but we disagree with her overall emphasis. While the aggregate employment effects of AI and robotics are uncertain and hotly debated, dismissal of automation as a 'charade' is deeply ahistorical. Generations of workers, from hand-loom weavers to assembly line auto-workers and cold metal print-setters would testify that there is nothing 'faux' about capital's tendency to replace humans with machines. The millions of people migrating from planetary zones bypassed by analog and digital supply chains and automated factories testify to the reality of surplus populations. While Federici may have been quite rightly suggesting that we should rethink who or what should be considered socially disposable, there is no doubt that capital always *has* made people and indeed entire populations 'disposable' (which, of course, is why it has to be resisted). Shrinking from that reality at the moment when a new instalment of corporate machinic power raises such disposability to a new level, and writing it all off as bluff and hype, may be reassuring, but it is unwise, sentimental and dangerously complacent. Probably recognizing this, at the conclusion of her essay, Taylor abruptly changes course, and concedes: 'There is no denying that technological possibilities that could hardly be imagined a generation ago now exist, and that artificial intelligence and advances in machine learning and vision put a whole new range of jobs at risk. Entire industries have already been automated into nonexistence.' And she rightly remarks that the 'emphasis on technological factors alone, as though "disruptive innovation" comes from nowhere or is as natural as a cool breeze, casts an air of blameless inevitability over something that has deep roots in class conflict' (2018). To which we say *d'accord*. But confronting these issues demands understanding AI and its automating capacities, accompanied though they are with abundant mystification and fetishization, as something more than just a 'charade'.

Our third object of critique, the left 'maximalist' position, is the diametric opposite of the 'minimalist' approach. Not only does it hold that AI and associated technologies, such as robotics, are 'for real', and have the capacity to drastically transform the conditions of production and work, it also sees these capacities as stepping stones to socialism. Proponents of this view look optimistically at the automating capacities of AI as an opportunity to ameliorate, perhaps eventually abolish, the exploitation of wage-labour, opening up prospects for a society in which people enjoy more free time, for pleasure, personal development and/or political engagement. This seems to offer a path for socialists that is more achievable than the daunting prospect of a full-scale revolution against capital. Instead, it can be attained by a social democratic government prepared both to foster the technologies of the fourth industrial revolution and to introduce a 'universal basic income' (UBI) or 'citizens' income' – a guaranteed payment to all citizens independent of any waged job. Lenin famously wrote that communism equals 'Soviets plus electrification'. It is fair to say that 'AI plus UBI' has become the formula for techno-progressive social democratic thought. A constellation of thinkers has formed around this attractor, articulated in works such as Nick Srnicek and Alex Williams's *Inventing the Future: Postcapitalism and a World Without Work* (2015); Paul Mason's *Postcapitalism* (2015) and Aaron Bastani's (2014, 2019) arguments for 'fully automated luxury communism'; the xenofeminist (Hester 2018) line of post-gender futurism; and a cluster of autonomist or *post-operaismo* theorists.

Again, we sympathize with and in many respects share the aspirations of this group; indeed, one of us has written about digital technologies in a very similar vein (Dyer-Witheford 2014), while another has suggested that Marxism might usefully incorporate similarly maximalist elements of transhumanist thought (Steinhoff 2014). However, we have written this book in part to directly challenge some premises of such AI-optimism. In particular, we want to contest the idea that AI can easily be detached, disentangled and re-appropriated from capitalism. Here it is useful to think about the sources on which the maximalist position draws – in part from Marx's own sometimes enthusiastic embrace of the modernizing powers of the forces of production to catalyse the emergence of socialism or communism, and also the uptake and reinterpretation of this position by poststructuralist theorists such as Gilles Deleuze and Félix Guattari. Perhaps the most important source, however, is the 'accelerationist' thinking of the anti-Marxist philosopher Nick Land (2012), who uncompromisingly argues for and celebrates what he sees as the unstoppable and species transforming (or terminating) power of computation. Indeed, the inaugurating document for the maximalist line of thought we have mapped here is Williams and Srnicek's 'Accelerationist Manifesto' (2013), which attempted a leftist re-do of Land's thought: the most accurate shorthand for the group of 'maximalist' theorists we have described is 'left accelerationists'.

As we will argue at length later, this appropriation of Landian thought dodges some of its originator's key arguments. For one, Land (2014) held AI to be the consummatory technology of capitalism, one that implanted the logic of capital at its very core. AI, in Land's view, is not merely appropriated by capital, but constituted by it: it is a technology made from and for its processes of labour automation, commodity acceleration and financial speculation. A second, yet more disquieting Landian point, is that this mutual embedment of capital and AI leads not to human emancipation from capitalism, but, on the contrary, to capital's emancipation from the human: a capital that no longer needs *homo sapiens*; human extinction.[1] These are not comfortable thoughts. And they are made even less comfortable by the fact that Land in his recent writings has emerged as a reactionary champion of racist and misogynist 'dark enlightenment' ideas that have in complex ways infiltrated the culture of Silicon Valley where much AI production takes place.[2] While we emphatically disassociate ourselves from this aspect of Landian thought, we nonetheless believe that any communist position on AI has to take his original accelerationist proposition – that AI has an elective affinity with capitalism and is fundamentally inhuman – seriously.

Given our anti-capitalist critique of both left minimalist denial and maximalist celebration of AI, it might be expected we go on to enunciate some middle-ground, moderate position. This is not the case, or true only in the sense that we want to remove the floor beneath both minimalist complacency and maximalist optimism. Our critique of AI can best be characterized as 'abyssal', and this in two senses. First, we confess, as we think other AI thinkers should, that there are vast indeterminacies about the directions and destinations of AI-infused capitalism. Peering into the conflicting estimates of AI's near and far future capacities and deployments can, and should, instil political vertigo. The everyday uses of AI now commonplace in advanced capitalism give some indicators of its future trajectory, but no certainties. This may seem an odd assertion for a Marxist theorization of AI, given that Marxism has in many incarnations asserted bold teleological certainties; however, as we argue later, Marx's work itself contains divergent accounts of the outcome of capital's technological compulsions. We read it as a matrix of possibilities, rather than a promissory note. In and of itself, this approach undercuts complacencies both that social struggles persist unchanged, regardless of new technologies, or, conversely that, because of the same new technologies, capital's self-destruction is imminent.

That said, however, the second 'abyssal' aspect of our AI analysis is that amongst the maze of future possibilities, some potential outcomes can be discerned that are far more deeply disturbing than is allowed by either the maximalist or minimalist positions, with their respective confidence about the continuation or the end of capitalism. These outcomes throw into question assumptions about the labour theory of value, the continued centrality of struggles at the point of production, or even the confidence that capitalism cannot survive the abolition of its human waged workforce. These points demand consideration, not to justify defeatism, but as a component of a revival of revolutionary communist thought. This is what we mean when we say that, at the same time as making a Marxist critique of AI, we make an AI-informed critique of Marxism. What then is AI?

'A MACHINE CAN BE MADE TO SIMULATE IT'

To understand the effervescence surrounding AI we need to define what AI is and how it functions. We are, emphatically, not AI experts; we will make errors deriving from our lack of technical knowledge as well as from the rapidly evolving nature of the field. Despite such difficulties, we believe grappling with basic AI concepts and how AI actually works is important. Too many accounts of AI, celebratory or dismissive, skip this effort. But it is only through some familiarity with the science and technology of AI that an effective critique can be mounted.

The workshop at Dartmouth College in Hanover, New Hampshire in 1956 is usually taken as the start of the field and study of 'artificial intelligence'. The organizers described their goal as follows:

> The study is to proceed on the basis of the conjecture that every aspect of learning or any other feature of intelligence can in principle be so precisely described that a machine can be made to simulate it. An attempt will be made to find how to make machines use language, form abstractions and concepts, solve kinds of problems now reserved for humans, and improve themselves. (McCarthy et al. 1955)

Since then, definitions of AI have been many and vague. AI experts show no consensus (Faggella 2018c). Compounding this definitional problem is the 'AI Effect', whereby as soon as AI can do something, it is no longer considered to require intelligence. Pamela McCorduck noted that in the history of AI 'every time somebody figured out how to make a computer do something – play good checkers, solve simple but relatively informal problems – there was chorus of critics to say, "that's not thinking"' (2004: 204). One recent AI textbook quotes Elaine Rich's pithy definition of AI from 1983: 'the study of how to make computers do things at which, at the moment, people are better' (quoted in Ertel 2018: 2). A more formal definition of AI we find useful is:

> The essence of AI – indeed, the essence of intelligence – is the ability to make appropriate generalizations in a timely fashion based on limited data. The broader the domain of application, the quicker conclusions are drawn with minimal information, the more intelligent the behaviour. (Kaplan 2016: 5–6)

This definition distinguishes AI from mere computation and allow us to differentiate between different types of existing and hypothetical AIs by considering their speed, quantity of information required, and generality of application. Kaplan's definition, however, says nothing about what AI looks like out in the world. AI does *not* mean robot, a confusion that can be blamed on pop culture. The roboticist Alan Winfield offers three complementary definitions of a robot:

1. an artificial device that can *sense* its environment and *purposefully act* on or in that environment;
2. an *embodied* artificial intelligence; or
3. a machine that can *autonomously* carry out useful work (2012: 8)

The most important aspect of Winfield's definitions is that, despite differing morphologies, all robots have bodies. AI, however, is software and, therefore, need not be embodied, though it requires computing hardware to run on. Advanced robots employ AI for functions including perception, planning actions, and learning, but a robot body does not necessarily entail AI, nor does an AI system necessarily entail a robot body.

To distinguish actually-existing AI from its speculative future incarnations, it is helpful to employ the following three categories: narrow AI, artificial general intelligence (AGI), and artificial superintelligence (ASI). Actually-existing AI is narrow: 'the vast majority of current AI approaches … are primarily designed to address narrow tasks' (Johnson et al. 2016: 4246). Most AI research, all commercial applications of AI, and the AI that consumers use daily, are such task-based tools. They are functionally more akin to microscopes than the anthropomorphic and politically active droid L3–37 in *Solo*. These systems have none or very little ability to do anything beyond their particular domain of functionality. An AI system that recognizes faces in photographs is not going to be able to process recordings of speech, play Go, or compose emails, and it is definitely not going to be able to speak Farsi. We will discuss dozens of existing narrow AI systems over the course of this book.

On the basis of generality, narrow AI is contrasted with artificial general intelligence (AGI), which refers to AI with 'the capacity for efficient cross-domain optimization' or 'the ability to transfer learning from one domain to other domains' (Muehlhauser 2013). AGI refers to an AI with the capacity to engage and behave intelligently in a wide variety of contexts and to apply knowledge learned in one context to novel situations, meaning it would be 'capable of reasoning across many intellectual domains' (Baum 2018a: 3). As of 2019, AGI remains a speculative technology, although serious research is now being conducted on it in both public and private institutions. We discuss AGI in Chapter 3.

Artificial superintelligence (ASI) is yet more speculative. While an ASI 'is likely to have general intelligence' (Baum 2018a: 3), it specifically refers to an AI 'that greatly outperform[s] the best current human minds across many very general cognitive domains' (Bostrom 2014: 63). ASI is a science fiction staple, but serious discussion of it also occurs in academic circles where it is often seen as swiftly following the creation of AGI (see e.g. Bostrom 2014; Torres 2018; Baum 2018a; 2018b). Most commonly, the scenario imagined is that an AGI gains the ability to self-modify and evolves into a god-like ASI with unpredictable powers. The consequences of such an event are impossible to predict with certainty, but the mere possibility of it occurring compels thinkers and institutions – including Nick Bostrom and the Future of Humanity Institute at Oxford, Seth D. Baum and the Global Catastrophic Risk Institute, as well as Eliezer Yudkowsky and the Machine Intelligence Research Institute (MIRI) – to argue that we must seriously research the possibility now.

Another important distinction is that between 'strong' and 'weak' AI. While sometimes the term strong AI is used to refer to AGI (Kurzweil 2005a: 260), the term originally derives from the work of the philosopher John Searle (1980), who used it to describe the position of those who believe that an advanced AI would be conscious. Searle critiques this view from the sceptical position of weak AI, which holds that machines can never be conscious. Searle's famous Chinese Room thought experiment, which hypothesizes a human (or machine) equipped with an encyclopaedic set of rules for translating Chinese into English without being able to speak or understand the language, attempted to prove this position. As Kaplan puts it, 'strong AI posits that machines do or ultimately will have minds, while weak AI asserts that they merely simulate, rather than duplicate, real intelligence' (2016: 68). We do not take a definite stance on the question of machine consciousness. The arguments in this book do not depend on machine consciousness being physically or even logically possible, nor on the impossibility of such. On this topic we are functionally agnostic, and, as we argue in Chapter 3, so too is capital.

Actually-existing narrow AI is typically divided into three schools of thought: Good Ol' Fashioned AI (GOFAI), machine learning (ML), and the situated, embodied and dynamical framework (SED). GOFAI, also known as symbolic AI, was the first approach to AI and remained dominant until the 1980s (Boden 2014: 89). It is an approach that aims to implement high-level cognitive functions, such as logical reasoning, in machines through the manipulation of information encoded in a symbolic language. Such a system creates internal representations of its world or a problem domain in a symbolic language and performs logical manipulations on this representation to think or act. These systems are often constructed out of sets of clearly defined rules. The best examples of GOFAI are so-called 'expert systems' or 'knowledge systems', which emerged and proliferated in the 1980s. These were intended to capture the knowledge of human experts and make it available to less skilled workers or ignorant managers. Expert systems were used for medical diagnosis, credit scoring and analysis, and business management, but the most famous is IBM's chess-playing system Deep Blue, which in 1997 defeated the reigning world champion Garry Kasparov. However, GOFAI required vast sets of rules with myriad possible interactions. Solving complex problems in this way necessitates tremendous computational power; for these and other reasons, approaches other than GOFAI were pursued.

One reaction to the problems of GOFAI is the 'situated, embodied, dynamical (SED) framework' (Beer 2014: 128); Rodney Brooks, a pioneer in the field, called his approach 'nouvelle AI' to emphasize its qualitative break with GOFAI (Copeland 2000). SED researchers are often motivated by 'Moravec's paradox' – the observation by roboticist Hans Moravec that 'it is comparatively easy to make computers exhibit adult level performance on intelligence tests or playing checkers, and difficult or impossible to give them the skills of a one-year-old when it comes to perception and mobility' (1988: 15). SED approaches to AI emphasize the irreducible importance of the body – with its perceptual apparatuses and morphology – to cognition: for this school, it is through solving material problems that machines can evolve intelligent behaviours. Such approaches are therefore often concerned with robotics and artificial life in addition to AI. SED practitioners initially focused on very simple, insect-like robots, but in the 2010s more complex, partially humanoid robots became possible and have been introduced into industrial settings. It is possible that some variety of the SED framework could be the next dominant AI paradigm.

Another reaction to GOFAI was machine learning (ML). The ML school, formerly called connectionism, existed as early as the Dartmouth workshop, gained some traction in the 1980s with advances in learning algorithms, but did not explode until the 2010s when big data and cheap computing power proliferated. As of 2019, ML is the dominant approach to AI.[3] It is a statistical pattern-recognition approach. One NVidia researcher has described ML as a process comprised of three steps: '(1) take some data, (2) train a model on that data, and (3) use the trained model to make predictions on new data' (Dettmers 2015). In other words, ML systems can be understood as creating their own models of inference.

While ML may operate on a variety of architectures, the cutting edge of AI in the 2010s has largely run on artificial neural networks (ANNs) – computer programs that are inspired by, albeit quite different from, the human brain. ANNs roughly mimic the electrical operations of the brain's neuronal connections rather than emulate high-level logic like GOFAI does. ANNs are 'based on the assumption that cognition emerges through the interactions of a large number of simple processing elements or units (i.e., 'neurons')' (Sun 2014: 109). Artificial neurons are organized into a series of layers and each layer is connected to the layers above and below. The lowest level receives inputs – e.g. images, text or speech – in the form of data that has been vectorised (converted into long strings of numbers). Higher levels – called hidden layers – process data that is sent up from the layers below them. Early networks had only one hidden layer, but today's networks have many more. In general, the more layers the underlying ANN architecture of the ML system has, the more complex patterns it can find. The most advanced ML in the 2010s is called 'deep learning' because these networks are many layers deep (LeCun, Bengio and Hinton 2015: 436–44), with some networks possessing as many as 1,000 (He et al. 2016). While previous ML networks were constructed by hand, a 'key aspect of deep learning is that these layers of features are not designed by human engineers: they are learned from data using a general-purpose learning procedure' (LeCun, Bengio and Hinton 2015: 436).

The artificial synapses which connect the layers of artificial neurons are 'weighted' with numeric values representing the strength of the connection. The network 'learns' through adjusting the weights of these connections. We can thus think of an ML system as a fixed template with changeable parameters; 'by assigning different values to these parameters, the program can do different things' (Alpaydin 2016: 24). ANNs are exposed to a data set, which might be images of faces or audio clips of people saying hello. In a process called 'training', the network is exposed to many instances of the chosen object(s) and the weights of the synapses are adjusted by a learning algorithm until the network learns to output the correct response, recognizing faces or the word hello, as the case may be (Kaplan 2016: 30).

It is useful to distinguish between the three broad types of machine learning – supervised, unsupervised and reinforcement learning. Supervised learning has been the most successful so far. In this approach, input data is labelled by human teachers, usually in terms of categories, and the system learns those categories by discerning patterns across the supplied examples. Given enough photos of red hexagonal signs with the word STOP on them, in various visibility conditions, and from various angles, a supervised learning system can learn the concept of a stop sign.

However, because of the necessity of labelling, supervised learning entails a lot of human labour. This has driven companies to develop techniques of unsupervised learning to enable a network to generate categories and labels on its own (Alpaydin 2016: 117). The idea is that with exposure to enough data, the system will identify 'incredibly sophisticated and complex correlations' across the data set (Kaplan 2016: 30). In so doing, it may be said to generate its own concept of a stop sign. Some deep learning pioneers argue this will eventually become the central ML approach because it mimics how humans and animals evolved to learn; not by being told what everything in the world is called, but through observing it (LeCun, Bengio and Hinton 2015: 442).

Reinforcement learning lies somewhere in between supervised and unsupervised learning. The pioneers of this approach describe it as 'learning what to do … so as to maximize a numerical reward signal. The learner is not told which actions to take … but instead must discover which actions yield the most reward by trying them' (Sutton and Barto 1998: 127). Reinforcement learning was thought to be limited to simple domains, until 2013 when the UK firm DeepMind combined it with unsupervised learning to teach a system to play Atari games with superhuman skill, without programming any knowledge about the games into the system, and giving it access only to the score of the games and the pixel information displayed on the screen (Knight 2017). The same combination enabled AlphaGo's win over Go master and world champion Lee Sedol.

As Kaplan emphasizes, the learning in ML should be understood as 'extract[ing] patterns from data' (2016: 27).[4] Instead of being built top-down as a set of rules for handling data, ML systems go bottom-up: 'learning algorithms … are algorithms that make other algorithms … computers [that] write their own programs, so we don't have to … [it is] the inverse of programming' (Domingos 2015: 6–7). This is why ML advocates see it as a Copernican revolution in programming. If a system can learn, the designer does not need to anticipate and program a solution for all potential situations the system is exposed to (Alpaydin 2016: 17). Ideally, the system will develop its own solutions and 'the data itself … defines what to do next' (Alpaydin 2016: 11).

This would not, however, be a very powerful attribute of ML if these systems could not generalize what they have learned to data not included in the training data set. Thus they are assessed according to their 'generalization ability' (Alpaydin 2016: 40). While the ML systems of 2019 are at best able to generalize to new data of a similar kind to that which they were trained on, increasingly sophisticated systems may approach AGI's hypothetical capacities for generalization across domains. This book is about the implications of AI, from machine learning to AGI, for the future of capitalism.

MACHINE MARX

Machinery is crucial in Marx's analysis of capital, so much so that almost any serious study of his work engages the topic in some way. Here we rapidly map those aspects of Marx's thought that are most important to the discussion of AI, highlighting points relevant for our arguments later in this book. There are, we suggest, three main strands in Marx's thoughts about machines. In the first, major line, the machine is a supplement to the human labour that is the crucial creator of value within the capitalist system. While production becomes increasingly machinic and intensifies the exploitation of workers, machinery ultimately contributes to the system's terminal crisis. The other two, minor lines, logically emerge from this major current but also depart from it. Both of them posit a moment at which the machine becomes autonomous from labour. However, one sees the consequences as liberatory, the other as nightmarish. These three strands of Marx's machine-thinking can, with care, be conjugated together, but this depends on glossing tensions between them that widen and deepen as we confront the conundrums of AI.

What we term the major line of Marx's machine analysis unfolds in the first volume *Capital*. In this account, machines, along with other equipment, buildings and raw materials, are 'constant, fixed' capital.[5] This is contrasted with the 'variable' capital of human labour (Marx 1990: 508–9; 1992: 237–48). This distinction between 'fixed' and 'variable' capital rests on the basic proposition that it is only human labour that creates value within capitalism: the machine, however gargantuan its powers seem relative to those of humans, can only act as a supplement or force-amplifier to the essential, human activity, increasing its efficiency, albeit by manifold times. Machinery, which has itself been built by humans, is 'dead labour'.

Rather than generating new value, the machine already has value which it transfers to the product: it 'yield[s] up its own value to the product it serves to beget' (Marx 1990: 509). The social function of fixed capital is to produce relative surplus-value, which it does by reducing necessary labour-time and, hence, increasing surplus labour-time, i.e. by 'shortening the part of the working day in which the worker works for himself, to lengthen the other part … he gives to the capitalist for nothing' (Marx 1990: 492). Increasing the productivity of labour means that the worker's output is increased: more commodities are produced in less time, and consequently these commodities are cheapened because less value is objectified in each individual commodity.

Marx's detailed explanation of machinery occurs in Chapter 15 of *Capital, Volume 1*, which concerns 'Machinery and Large-Scale Industry' (1990: 492–639). Here, Marx was concerned 'only with broad and general characteristics' of machinery (1990: 492). The genealogy of machinery is found in tools; the instruments of handicraft labour are turned into machinery, automation technology being 'fully developed machinery', which has three different parts: (1) the motor mechanism that is the driving force or motive power; (2) the transmitting mechanism that regulates, changes and distributes motion; and (3) the tool or working machine that uses this motion to modify the object of labour (1990: 494). A key component of any machine is its 'emancipation from the restraints of human strength' which occurs when it obtains a regular and controllable motive power; Watt's double-acting steam engine being the first major 'self-acting prime mover' (1990: 502). Whereas with mere tools the process of production is adapted to the worker, the system of machinery is a 'vast automaton' which confronts the worker as a 'pre-existing material condition of production' (1990: 508).

The section on 'The Struggle Between Worker and Machine' in Chapter 15 emphasizes how capital's technological dynamic is inseparable from class conflict. Marx examined the paradox by which, under capital, labour-reducing machinery creates a hell for labourers: the mechanical lightening of demands for physical strength catalyses the large-scale induction of women and children into factories; the capacity of machines to run indefinitely, and the need to pay for their purchase, leads to a prolongation of the working day; the ability to accelerate and multiply machine operations results in the intensification of work. Marx also addressed technological unemployment, describing how '[t]he instrument of labor strikes down the laborer' in a process where machinery 'act[s] as a superior competitor to the worker, always on the point of making him superfluous, and capital proclaims this fact loudly and deliberately, as well as making use of it' (1990: 562). Indeed, Marx suggested 'it would be possible to write a whole history of the innovations made since 1830 for the sole purpose of providing capital with weapons against working class revolt' (1990: 563).

Mechanization is also propelled by competition between rival capitalists. The value of a commodity depends on the amount of socially necessary labour expended in its production. If a capitalist can, by introducing technology, reduce the labour for which she pays, while still selling the product at the prevailing price, she will enjoy greater profits than her competitors. This advantage will eventually be neutralized as use of the labour-saving innovation becomes generalized, but this just sets the scene for the next wave of automation. Thus both class conflict and competition between enterprises give capital an intrinsic drive to replace humans with machines. Marx described this as the tendency for capital to increase its 'organic composition', that is to say, the proportion of constant (machines, buildings and raw materials) compared to variable capital (labour) (1990: 762).

Machinery also, however, throws capital into crisis. In *Capital*, Marx alternated between two explanations as to why this is. One, perhaps the most readily understandable, is that because machinery enables capitalists to increase production while (other things being equal) reducing their wage bill, it fosters gigantic imbalances between the increasing volumes of commodities produced and the purchasing power available to buy them. This brings on economic stagnation and paralysis, with factories closing and unemployment lines growing, until enough firms go out of business to eliminate the glut of overproduction and get the system moving again – or a social revolution breaks out.

Marx's other explanation, less intuitively obvious, but arguably more profound, has to do with the tendency to a falling rate of profit (FROP) (1991: 317–38). Because the value of a commodity ultimately depends on the amount of socially necessary labour required for its production, replacing humans with machines lowers the value of the commodity – and hence, eventually, the price it commands, and the profit per item capital can command. Because automation cheapens goods, it makes each single one of these goods less profitable to the capitalist. In the face of this tendency of the *rate* of profit to fall – a direct result of automation – capital can, at least for a time, maintain the *mass* of profits by increasing the sheer volume of production, but it is running against its own value-decreasing machinic momentum. Marx (1991: 339–48) listed a number of ways capital can hold this process at bay or even temporarily reverse it, but the tendency of increasingly machinic production is, again, towards spasmodic crisis, this time caused by the flagging profitability of business, and moments of unemployment, immiseration and social tumult.

These two versions of crisis theory, and their degree of compatibility, have been intensely debated. But in both versions, capital's recurrent crises arise from its inherent drive to substitute constant (machinery) for variable capital (labour). This either reduces consumption power (by cutting wages) or lowers the profitability of production (by cheapening goods), or both. The outcome is repeated throughout deepening cycles of economic breakdown, each of which offers the possibility of a revolution by a working class suffering from downward pressures on wages and the threat of unemployment, but still central enough to production to halt or take control over it. This, then, with its internal bifurcations and attendant controversies, is what we call the major line of Marxist thinking about capitalism and machines. An analysis of AI undertaken within this current will analyse it in terms of labour exploitation, inter-capitalist competition and capitalism's techno-induced crisis tendencies.

What we dub the two minor tendencies in Marx's thought are extensions of, but also deviations from, this major line. Latent in Marx's account of capital's increasing mechanization is the idea that the positions of worker (initially the main, value creating actor) and the machine (at first the worker's power-amplifying supplement) invert. The worker, who at the handicraft stage was the subject of the labour process, becomes an automaton of repetitive, repeated motions, responding to automatic machinery rather than using it; the automatic machinery has become the subject of the labour process. Two subsidiary, perhaps maverick, lines of Marx's thought develop this idea to its logical conclusion – but in two different directions.

The first, and by far most famous, comes from the 'Fragment on Machines' in Marx's *Grundrisse*, which from the 1970s on has been seen as an extraordinary anticipation of high-technology capital. In the 'Fragment', Marx envisaged capital making vast techno-scientific achievements by mobilizing the 'general intellect'. This enables it to reach a level of automation that, while not eliminating human labour entirely, reduces and relegates it to the peripheral position of supervising a mainly machinic process. This might seem the final triumph of capital over its troublesome working class, but Marx in the 'Fragment' presented it as a pyrrhic victory. By removing the necessity to base production on wage-labour (and hence liquidating the possibility of basing consumption on waged income), it undermines value, i.e. the whole basis of capital's social organization. Automation inadvertently subverts capital by abolishing work. This is consonant with other celebrated passages of Marx's concerning the liberatory nature of technological development, most notably the account of how developing 'forces of production' burst apart fossilized 'relations of production', making way for the appearance of a whole new 'mode of production' (Marx 1973 [1859]).

At first glance, the 'Fragment' might seem just to restate in especially emphatic form the predictions about the mounting organic composition of capital that inform Marx's thinking about the FROP. As George Caffentzis (2013) points out, however, there is a divergence between the two theories, for while the FROP depends for its operation on the validity of the labour theory of value, the 'Fragment', in contrast, posits a dynamic of capitalist collapse arising from the liquidation of labour – and value. The *Grundrisse* was translated into English, French and Italian at the very time when computers were first beginning to enter the workplace, and from the moment of its appearance it was taken as harbinger of a high-technology, 'cyborg' communism which would overcome all the difficulties a drab, industrial, actually-existing socialism was experiencing in organizing work on a non-capitalist basis by simply eliminating the need to work at all. It is therefore no surprise that the 'Fragment on Machines' is a foundational document for theorists of left accelerationism, postcapitalism and fully automated luxury communism, so much so that Frederick Harry Pitts (2017) has dubbed these all instances of 'Fragment Thinking'.

A second minor strain in Marx's thought, less discussed than the 'Fragment', is the nightmare vision of capital in the 'Results of the Immediate Process of Production' (appearing in some editions of *Capital* as an Appendix to Volume 1) (1990: 949–1084). In this text, Marx gives an account of the process of capitalist 'subsumption' – roughly translated as domination, envelopment or take-over – of labour. He details two moments: formal and real. In the first moment, capital organizes labour as wage-labour, thus merely changing its social form, while leaving the content of labour, i.e. how it is carried out, the same as in pre-capitalist artisanal handicraft. In the second moment, however, the content of labour changes – it is really subsumed – to better meet the dictate and demands of the capitalist production of surplus-value. Initially, real subsumption occurs through introducing a division of labour into the handicrafts, but subsequently by the automation of labour by machinery, which requires capital to absorb socially produced scientific knowledge to develop technologies adequate to and commensurate with its own priorities – notably, the automation of production and the acceleration of the circulation of commodities. In the transition from formal to real subsumption, 'absolute' exploitation of labour – the extension of the working day – is displaced by the extraction of 'relative surplus-value', which is based on increasing the productivity of labour by intensifying the labour process through the division of labour and machinery. As this process builds, Marx argues that a situation emerges in which, in the industrial factory (only nascently visible in his time), the worker confronts a fully 'alien power' that appears endowed with a 'colossal independence' from human agency, rendered 'autonomous' by techno-science.

This account might seem just another description of the mounting organic composition of capital, or indeed of the semi-automated 'animated monster' of capitalist machinery featured in *Grundrisse* (1993: 470). There are, however, differences in inflection. For one thing, since the subsumption argument is that capital actually adopts machinery that it designs to its systemic requirements (the valorization of value through elimination of human labour and acceleration of commodity circulation), it becomes more difficult to envisage how the 'forces of production' conflict with 'relations of production' – if anything, the former would seem to reinforce the latter. There is certainly no hint in the 'Results' of either the crisis-inducing falling rate of profit or of the self-destructive labour-abolishing logic of the 'Fragment'. There is just the towering presence of an all-but-incomprehensible production apparatus that looms over and surpasses the worker it once depended on. This is all the more apparent because, while the first half of the appendix includes one of the most complete discussions in Marx of the powers of the 'collective worker' engaged in the cooperative fusion of various types of work, from manual labour to engineering, by the end the collective worker is dwarfed and seems virtually obliterated by the machine apparatus to which its powers have been transferred.

It can fairly be argued that, read as an appendix to *Capital*, this document should be understood only in the light of what precedes it, so that capitalist breakdown can be assumed. And it can further be asserted that the machinic autonomy of capital is only an 'appearance' – a mystification that has to be seen through to detect the continuing, if baroquely veiled, importance of labour-power situated down remote production chains. Maybe so. But as even chronic optimists like Antonio Negri (2017) note, 'appearance' in Marx doesn't mean 'shadowy, superficial or insubstantial': on the contrary, it means a concrete social reality created on the basis of a mystified and disguised process, namely the incremental subsumption of the power of workers into machinery. While Marx developed this sombre vision through analysis of the industrial factory, it invites thought about a further stage of 'hyper-subsumption' in which capital's autonomizing force manifests as AI: later in this book we explain how this new stage of subsumption is unfolding, and may culminate in AGI.

Both the major and minor lines of Marx's machine analysis describe dynamics in play in the actually-existing AI-capitalism of 2019. We are not at the heights or depths of machinic capitalism foreseen in either the 'Fragment' or the 'Results'. Workers on Foxconn assembly lines, in Amazon warehouses or on Facebook content moderation sites all attest to the continued use of machines to intensify and speed up human labour. Crises such as the 2008 Wall Street crash reveal the deep instabilities which machinically depressed wages, digitally organized cheap labour and high-speed financial trading are bringing to capitalism. So much of the major line of Marx's machine thought holds true, perhaps truer than ever, as the 2020s approach. But ML-driven AI, developed in part in response to the crisis of capitalist globalization, is placing on the horizon possibilities that resemble those in Marx's visions of capitalism's machinic extremes. The 'Fragment on Machines' and the 'Results of the Immediate Process of Production' are recto and verso of one page, a page that speaks both of machine power liberating humanity from capital, and of a capital rendered autonomous from humanity. Capitalists and communists alike, be careful what you wish for!

IN THE AGE OF SELF-REPLICATING AUTOMATA

Marx's discussion of machinery stops with the steam-powered factory. Only towards the end of his life were electricity and electromagnetism harnessed, and although early computer technology like Jacquard's Loom and Babbage's failed Difference and Analytical Engines existed when he wrote, he did not discuss them. So to bring Marxism to bear on AI, Marx's account must be amended. There is a formidable Marxian literature on 'cybernetic capitalism' and 'digital labour', but Marxian analysis specifically devoted to AI, analysing the specificity of its technology, political economy and class implications, is relatively rare. There are, however, three treatments that have influenced our thinking.

The first is Tessa Morris-Suzuki's examination of Japan's high-technology capitalism in the 1980s, the period which saw the introduction of robots in auto manufacture, the explosive growth of the video game industry, and projects such as Japan's ultimately doomed 'Fifth Generation Computer Systems', an early AI initiative supported by Japan's Ministry of International Trade and Industry. Writing in this context, Morris-Suzuki suggested – much as we have here – that some of the disarray of the contemporary left stems from its reluctance to confront the possibility of a highly automated capitalism, instead taking 'one of two contrasting positions' – either denial 'that the contemporary "information revolution" represents any fundamental change in the nature of capitalism' or the assertion 'that it spells the death agony of the capitalist system' (Morris-Suzuki 1986: 81). Taking issue with the claim by the famous Marxist scholar Ernest Mandel (1975: 207) that large-scale automation of production constitutes the 'absolute inner limit' of capitalism, Morris-Suzuki argued it was time to consider ways capital might perpetuate itself under such conditions. These, she said, would include the transfer of labour from production to 'perpetual innovation', a proletarianization of technical jobs, the corporatization of an education system geared to the production of elite research scientists, and the creation of a workforce 'easily taken up and easily discarded' (Morris-Suzuki 1984: 120). This now reads as a prescient description of the present. The new wave of AI poses the same problem Morris-Suzuki articulated, but at a higher level. ML and other new AI techniques are beginning to encroach on the activities she saw as the only available refuge for labour chased out of industrial production by machines.

The second is Ramin Ramtin's remarkable *Capitalism and Automation: Revolution in Technology and Capitalist Breakdown* (1991). In this book, Ramtin made the first systematic attempt to rethink Marx's theory of machinery in the light of the cybernetic technology that was driving the digital automation of the 1980s. He proposed that to Marx's three-part anatomization of industrial machinery, comprising motor power, transmission mechanism and tool head, had to be added the guiding or control function – a function once considered dependent on human intelligence and senses but now increasingly automated with information, including sensor, technology. By proposing this revision Ramtin offered a way towards a theorization of computers that, without endorsing the post-industrial euphoria about the information revolution, also recognized the qualitative change digital technologies brought. Ramtin's work was, however, also notable for the unflinching eye it cast on the possible consequences of this development for a Marxist analysis of class struggle. He suggested that the full-scale cybernetic onslaught of capital against its working class would bring to the fore issues of unemployment that had receded into the background during the postwar boom. His insistence that Marx's notion of proletarianization be recognized as a concept not just of workplace exploitation but also of the liability of ejection from work in many ways anticipates the discussions of 'surplus populations' that would emerge in the wake of the 2008 crash, which we take up later in this book.

Morris-Suzuki and Ramtin not only pointed to important changes in work and labour conditions associated with early AI. Between them, they also provided important revisions of Marx's basic conceptualization of machinery. Ramtin, drawing on cybernetics theory, pointed out that Marx's account of machines omits the key function of the control, presumably assuming it is ultimately directed by a human agent. This assumption is, however, Ramtin pointed out, precisely what was challenged by cyberneticists as they introduced the theory of feedback into machine operation. It is, he suggested, precisely the control function that distinguishes automation from mechanization and makes it qualitatively new. Morris-Suzuki emphasized the vastly increased scope and flexibility of machine application that comes with the separation of 'hardware' and 'software': with machinery whose operations can be changed by the switching of instructional programs, so that machines start to attain some of the variability that had previously been seen as unique to human labour.

While Ramtin and Morris-Suzuki provide analytic anticipations of AI drawn from early moments of digital automation, our third exemplar of Marxist AI analysis, George Caffentzis, provides a crucial theorization as to why many of the predictions of jobless futures of that era have not come true. In a series of essays written from 1980 to 2008,

Caffentzis argued that the apparent job-destroying powers of AI had to be considered in light of its antithesis, the expansion of the service sector and global sweatshops; one had to think 'Africa' and 'automata' together. He draws on the ninth chapter of Volume 3 of *Capital*, 'Formation of a General Rate of Profit (Average Rate of Profit) and Transformation of Commodity Values into Prices of Production' (1991: 254–73). Here Marx suggested that while the profit extracted by capital as a whole depends on the overall amount of surplus-value extracted within its entire system, there is no direct correspondence between any individual capitalist's profit and the amount of socially necessary labour they employ. Value is a social phenomenon and any and all value produced goes into a social pool after the commodities have been exchanged. But capitalists also appropriate surplus-value from this pool in the form of profit; an individual capitalist will appropriate more profit if their capital, relative to other capitals, is of a larger size, has a higher organic composition, and a higher average profit rate (Marx 1991: 241–73; Caffentzis 2013: 132–4). Thus highly automated businesses syphon-off the surplus-value generated by labour-intensive capital. Caffentzis called this 'the law of the increasing dispersion of organic composition', by which 'every increase in the introduction of science and technology ... in one branch of industry ... will lead to an equivalent increase in the introduction of low organic composition production in [an]other' (2008: 65). Caffentzis's account of capital's contradictory movements towards high-technology (automata) and low-wage-labour (Africa) suggests reflection on how its current AI-fever is induced not only by technological breakthroughs, but by increasing frustrations in finding cheap labour.

AI NOVUM

Before and during the writing of this book, in addition to assimilating materials about Marxist theory and the science, economics and sociology of AI, we read and watched a great deal of AI science fiction (henceforward, AI-SF), including not just the classic films – *2001*, *Blade Runner*, *Terminator* – that are inevitable points of reference for all AI discussion, but a wave of more recent writings and productions that accompanied the emergence of ML. Some will regard this as evidence of impaired judgement. Harry Collins (2018: 5–13) renames AI as 'Artifictional Intelligence' because, he argues, SF has encouraged a widespread overestimate of AI capabilities: depictions of the purported superhuman singularity encourage 'the surrender' – human abdication to stupid computer programs. Similarly, Mike Cook (2018), an AI researcher, designates AI-SF, alongside excessive respect for scientists and the over-selling of AI, as one of a set of cultural factors contributing to a 'basic lack of understanding' about what ML can and cannot do.

We agree, but think that this is not all there is to AI-SF. Darko Suvin (1979) famously proposed that SF cognitively explores potentialities incipient in a society at a given moment by focusing on a 'novum' (Latin for 'new thing') – a term he borrows from the Marxist theorist Ernst Bloch (1986 [1955]) to denote some new force appearing on the 'front line of historical process' (Suvin 1979: 63–84). Centring itself on the novum of AI, and thereby estranging and de-familiarizing our current reality, AI-SF conducts thought experiments about the possible directions – dizzyingly utopian, terrifyingly dystopian or depressingly mundane – of actually-existing AI-capitalism. In the case of AI-SF, the importance of these thought experiments is accentuated because of the now well-documented feedback loop between computer science and science fiction, in which scientists inspired by SF work towards the actualization of its imagined worlds.

It is, of course, quite true that there is not much thought behind many AI-SF thought experiments. Collins is properly scathing about Hollywood representations of AI as a superhuman James Bond villain or sexual manipulator. Such anthropocentric depictions obscure the profoundly 'inhuman' nature of AI. Techno-amplifying already familiar cultural industry tropes, they are complicit with corporate promises that AI offers us a future the same as the present, only bigger and better. However, there are also other kinds of AI-SF that are far more critical in their perspective. While composing this book we have sometimes tried to categorize such works according to which of the contending left perspectives – sceptical, accelerationist and abyssal – they might correspond to.

Cyberpunk AI-SF has affinities with the sceptical Marxist perspective on AI, even while it extrapolates technological capacities well beyond what the sceptics would consider plausible. This is because cyberpunk fictions give an unsparing anatomization of capital in an era of intelligent machines. They defamiliarize the squalors of actually-existing AI-capital by projection into a future where truly shining AI fuses with grimy proletarianization to yield a glistening *noir*. Superbly realized in the original *Blade Runner* (though not so much in its visually impressive yet oppressively patriarchal *Blade Runner 2049* remake), this genre continues to be embellished in recent AI-SF. Judd Trichter's novel *Love in the Age of Mechanical Reproduction* (2015) sends its hero on a grotesquely picaresque journey across a decaying Los Angeles where androids and humans coexist in a state of mounting antagonism and illicit liaison, seeking to recover the parts of his disassembled android lover: capital deepens its organic composition while relentlessly maintaining its domination over its fixed and variable components alike. Different in tone and pace, and original in its feminist thematics and domestic settings, yet sharing a similar problematic, is Andromeda Romano-Lax's *Plum Rains* (2018). Set in 2029 Japan it deals with a confrontation between a migrant Filipina worker and an intelligent machine over the care of elderly women, opening onto a complex and melancholy meditation on precarity, colonization and slavery. Although works of this sort may end on hopeful prospects of combined android-human revolt or escape, their overall tenor is to emphasize the subjugation of AI to the brutal equation by which capital owns machines, exploits humans, and substitutes one for the other as profit dictates.

A whole other species of AI-SF is, however, a haven of socialist utopianism. The possibilities of AI-based social and ecological planning underpin many of Kim Stanley Robinson's explorations of postcapitalist futures, most explicitly in his *2312*, where one of the forces enabling a break with an old earth 'decisively under the thumb of late capitalism' by other planetary settlements is the possibility that 'the total annual economy of the solar system could be called out on a quantum computer in less than a second', so that 'supercomputers and artificial intelligences ... make it possible to fully compute a non-market society' (Robinson 2013: 125). However, the most striking example of this genre is the late Ian Banks's great series of 'Culture' novels, from *Consider Phlebas* (1988) to *The Hydrogen Sonata* (2013). In the Culture hyper-intelligent and benevolent 'Minds' – evolved AIs – preside over a galactic commonwealth, coexisting with a humanity that, relieved of material need and even mortality by machinic tutelage, continues an adventurous, individualized and complex unfolding of its species-being. Here, in post-scarcity society, planning has been rendered redundant by plenitude – a fictional rendition of 'fully automated luxury communism' (Merchant 2015).

Diametrically opposite to such optimistic visions, and closer to our own abyssal perspective, is a series of depictions of AI as a 'bad novum' – not because of any anthropocentric enmity, but rather as the systemic culmination of runaway capitalism. The classic example is Charles

Stross's *Accelerando* (2005), in which what initially seems a light-hearted tale of digital entrepreneurialism turns into a harrowing story of how AIs proceed to dismantle the solar system as raw material for their ever expanding computational, marketized network, eating the universe in a competitive race that casually discards hominids as sub-optimal, under-performing agents. What *Accelerando* proposes is that, contra the ecstatic visions of Kurzweil and other utopian prophets of the Singularity, the real meaning of such an event is likely to be that 'The destiny of intelligent tool using life [i]s to be a stepping stone in the evolution of corporate instruments' (Stross 2005: 240; see Shaviro 2009). A somewhat similar recent vision comes from computer scientist and AI researcher Zachary Mason (who has worked on Amazon's ML-powered recommendation systems) (Locke 2017). In his *Void Star* (2017) protagonists in a San Francisco beset by inequality and climate change find their personal lives effectively subsumed and assimilated by the activities of indecipherable superintelligent AIs. These AIs 'engage with humanity only as byproducts affected by their actions, while they compute otherworldly questions of symbol manipulation' (Locke 2017). Massive and eerie derangements of personal identity result as a billionaire plutocrat attempts to mobilize such AI instrumentally.

The truly great AI-SF is, however, perhaps that which slips across, plays with and permutates the possibilities we have schematically reviewed. Still unsurpassed in this regard (and of special interest because of the author's evident deep engagement with Trotskyism) is Ken MacLeod's extraordinary *Fall Revolution* quartet (2008; 2009), which, by adopting a 'branching futures' conceit, comprehends a series of AI social outcomes ranging from human-controlled AI planned-economies to hive-mind digital absorption to a de-growth abjuration of AI. A rather similar effect is achieved by the sequencing of Peter Watts's recent *Sunflowers* short story and novella (2018) cycle. In this sequence, the protagonist's early ecstatic connection to AI yields to cynicism and then horror as she is recruited in an endless interstellar exploration journey under the direction of a narrow AI – 'the chimp' (Watts 2018) – because the corporate directors of this colonizing journey dare not instantiate AGI that might escape their control. For this reason, the cargo includes a cryogenically frozen crew summoned out of sleep at century-long intervals whenever an emergency requiring lateral thinking occurs. Yet despite this function, the humans are, it transpires, also expendable on a strict cost-benefit basis. The most recent story in this series ends with the defeat by AI surveillance powers of an attempted human mutiny. At the close, the heroine, who has made herself complicit with the suppression to forestall the oxygen-starvation death of all her co-workers, reflects that the emergence of an autonomous AI might offer the best possibility for the crew's escape from enslavement. Yet the overall sombre tone of Watts's universe raises the obvious question of whether such an intelligence would be an ally or an enemy to its proletarian fabricators.

In such instances – that is to say, at its best, rather than its worst – SF is a machine for thinking, and in the case of AI-SF, a machine for thinking about machine thinking and capitalism. For that reason, in the following chapters we occasionally weave references to AI-SF into our analysis.

CHAPTER SUMMARY

In what follows we draw on the work of Marx and Marxist scholars to make our own assessment of the present state and future prospects of capital's rendezvous with AI.

Chapter 1 presents a political economic account of the current state of what we term the AI industry. It describes the main protagonists – the giant tech companies in the US, China and elsewhere, and their interaction with both state research programmes and communities of open-source AI developers. It then goes on from current analysis to near future prognostication, suggesting that the ambitions of the great AI corporations point towards the establishment of AI as a new component of 'the general conditions of production', as a ubiquitous infrastructure, akin to the railways of the first industrial revolution or the electrical utilities of the second, on which all other forms of commodity production and circulation will come to rely.

Chapter 2 takes up the issue of AI and employment from the perspective of autonomist Marxism's class composition theory. It argues that AI should be seen as a second wave of a cybernetic offensive waged by digital capital against its working class, a new onslaught occasioned by the 2008 economic crisis. After reviewing some applications of AI within the social factory of advanced capitalism, we review the debate amongst futurists and economists as to whether AI will generate an imminent employment apocalypse or just a continuation of capital's processes of job destruction and creation. Whether or not AI brings about an immediate jobs crisis, many other aspects of its deployment are likely to exert downward pressure on wages and working conditions, and it is already precipitating an array of social struggles in and beyond the workplace.

Chapter 3 challenges the (to some) reassuring assumption that capital could not survive omnipresent AI automation. Taking its orientation from value theory, and assessing the long-term possibilities of AGI, it proposes that the humanist assumptions underpinning this belief no longer hold; *Homo sapiens* is not necessarily the only possible subject of capitalist proletarianization. If AI approaches or attains the horizon of singularity, the vistas that open up are not therefore those of inevitable capitalist collapse, but rather of the elevation of machine capital as a literally automatic subject autonomous from human beings. Social democratic programmes of 'full automation now' may therefore merely be positioning themselves as benign accomplices to this trajectory.

Our Conclusion draws out some political assumptions of the preceding analysis. While uncertainty is inescapable in thinking about AI, socialist strategies for reforming AI-capital by introducing a universal basic income and eco-modern techno-planning fail to confront the depth of the problem AI presents to projects of human emancipation. A communist orientation to AI focuses on transforming the ownership of the means of production so that real choices can be made about the adoption or abandonment of such technologies. The emergence of a new mode of production is, moreover, likely to occur under conditions of extreme social conflict and ecological disaster. In this context, it is not only capitalism that will be inhuman, for the form of the human that emerges, if any does, from the struggle against AI-capital will not be the same as that which entered into it.

1
Means of Cognition

[T]he Microsoft view is that AI needs to be included – or in Microsoft speak, 'infused' – in everything, from a simple word processor to a quantum computer.

James Thompson (2018)

THE NEW ELECTRICITY

In 2016, Andrew Ng, Stanford professor, entrepreneur, former Chief Scientist at Baidu and former head of Google Brain, pronounced AI 'the new electricity' and argued: 'Just as electricity transformed almost everything 100 years ago, today I actually have a hard time thinking of an industry that I don't think AI will transform in the next several years' (Lynch 2017). Ng is not the only one to espouse the notion of AI as a basic utility leading to a new industrial revolution – the idea is implicit in proclamations of a 'fourth industrial revolution' issued by capitalist institutions such as the World Economic Forum (Schwab 2017). It has also been explicitly advanced by tech guru Kevin Kelly (2014), who predicts that in the near future we will have a 'common utility' of 'cheap, reliable, industrial-grade digital smartness running behind everything ... Like all utilities, AI will be supremely boring, even as it transforms the Internet, the global economy, and civilization. It will enliven inert objects, much as electricity did more than a century ago. Everything that we formerly electrified we will now cognitize.' Companies such as Viv (n.d.) deploy this idea in their business plans, asserting that with their AI platform 'intelligence becomes a utility'.

Predictions such as those of Ng and Kelly suggest that AI could become part of what Marx referred to as the 'general conditions of production' (Marx 1990: 506; 1993: 530), i.e. the technologies, institutions and practices which form the environment for capitalist production in a given place and time. Marx spoke of infrastructure, which includes the means of communication and transport, as a significant component of the general conditions of production. If AI becomes the new electricity, it will be applied not only as an intensified form of workplace automation, but also as a basis for a deep and extensive infrastructural reorganization of the capitalist economy as such. This ubiquity of AI would mean that it would not take the form of particular tools deployed by individual capitalists, but, like electricity and telecommunications are today, it would be infrastructure – the means of cognition – presupposed by the production processes of any and all capitalist enterprises. As such, it would be a general condition of production. We propose the term 'means of cognition' – the AI-equivalent to Marx's means of communication and transport – but insist that it not be conflated with the *post-operaismo* notion of 'cognitive capitalism' (Moulier-Boutang 2011), for reasons we discuss in the conclusion of this chapter.

To make this argument first requires a review of the history of capitalism's adoption of AI, a survey of some existing and anticipated commercial applications of AI founded on the ML approach, and analysis of the contemporary AI industry. While the basis for accumulation in this industry is a highly advanced techno-scientific commodity, it is, like all capitalist enterprises, governed by compulsions to produce surplus-value, i.e. seek profit, compete, attract investment, control markets and defeat rivals through the formation of oligopolies and monopolies. We draw attention, however, to two structural features of this industry that could contribute to AI becoming a part of the general conditions of production: its lavish support and subsidization by neoliberal nation states eager to foster AI development for economic, administrative and military purposes; and the seemingly anomalous presence of a large and vigorous open-source component to AI research, in which tools and templates are distributed for free and worked on cooperatively, but are nevertheless channelled towards the platforms and priorities of AI oligopolists.

We speculate on how, in the near future, ML-enabled functions of cognition and perception could become ubiquitous via applications ranging from simple chatbots up to smart cities and the Internet of Things (IoT). These examples demonstrate some ways AI could be positioned as a general condition of production. This analysis paints a picture which runs counter to *post-operaismo*'s humanist reconfiguration of the notion of the 'general intellect' (Marx 1993: 706) as referring to the novel capacities of a networked multitude. Contrarily, the possible future of AI as part of the general conditions of production supports Marx's original formulation of the general intellect as capital's accumulated machinic capacities, *excised from* social human labour. While AI development does, for the moment, depend largely on the mining and processing of data drawn from a networked multitude, the aim of such development is to attain a whole new level of automation giving capital unprecedented independence from labour.

THE AI INDUSTRY AND THE OLIGOPOLISTS OF MACHINE INTELLIGENCE

While corporate interest in the actual and potential uses of the new AI are manifold, ranging from retail sales to entertainment and industrial production, the actual production of AI systems is a central concern for a more limited circle of high-tech companies. We refer to this complex as 'the AI industry', distinct from the broader field of commercial AI applications. While business-oriented publications continually remind us that AI will 'revolutionize' capitalist production (Columbus 2016), our analysis suggests that such a transformation, if it occurs, is still in its earliest phases. Instead, we see AI as one emerging industry whose influence is tied up with that of other emerging technologies and is as yet difficult to ascertain with certainty. Although business interest in AI is high, outside the AI industry this does not entail high levels of actual investment in the technology. A 2017 survey of attendees at an applied artificial intelligence conference concluded that 'AI adoption … remains low with the majority of major success stories coming only from the largest tech players in the industry' (Rayo 2018).

AI development first appeared as a distinct industrial sector in the 1980s. This first era of the AI industry was based around GOFAI expert systems. During this era, the AI industry consisted of a few small companies which produced systems as means of production for, and typically in cooperation with, their corporate customers. In some cases, large firms established internal AI departments to develop proprietary expert systems. Such systems required a considerable degree of specialization, had extremely narrow fields of application and required a lot of labour to produce and update. While attempts were made to develop 'generic' expert systems which could be applied to any field, they ultimately failed (Roland and Shiman 2002: 205). The commercial craze for these systems subsided in the 1990s, but around the same time the ML approach gained traction in academia and, during the 2010s, returned AI to the commercial realm, propelled by advances in computing power and improved learning algorithms. By 2017, *The Economist* (2017a) was proposing a shortlist of domains in which ML's power to 'sift through data to recognize patterns and make predictions without being explicitly programmed to do so' was becoming commercially important. It is worth surveying a few.

The ML-based AI industry is much more diverse than the first era of expert systems; this is one reason why advanced capitalism has recently contracted a serious bout of AI fever. *The Economist* (2017a) has been enthusiastic about the prospects of targeting online advertisements and product recommendations; the creation of virtual personal assistants and of augmented reality systems; and autonomous vehicles. As of early 2019, some of these were already highly advanced, while others only incipient. Algorithmic targeting of advertisements and recommendations has been a foundation of digital Web 2.0 enterprises for over a decade. Digital personal assistants, such as Apple's Siri, Amazon's Alexa or Microsoft's Cortana, are gradually becoming commonplace. Augmented reality (AR) products, overlaying physical reality with a mesh of virtual images and information, are only beginning to be sold as commodities or distributed as free vehicles for in-app purchases and data mining. Games such as Pokémon Go and other mobile apps are testing the AR waters, while further frontiers, such as medical applications, are being actively researched by companies such as Google, Apple and Microsoft. Perhaps the biggest prize for the commercial use of ML, but also its most daunting challenge, is the creation of self-driving cars and trucks, a 'moonshot' that has attracted leading information companies such as Google and Baidu, established auto-industry giants such as Ford, General Motors and Daimler, and upstart entrants such as Uber and Tesla, all racing to transform capitalism's entire transportation sector.[1]

AI industry enterprises build ML technologies, often initially for use in their own business operations, but also as commodities for sale or rent, or as a 'free' service. They produce commodities for both of the major 'departments' into which Marx divided society's total product and its total production process: (Department 1) means of production, i.e. commodities intended for productive consumption; (Department 2) means of subsistence, i.e. commodities destined for individual consumption (1992: 471). Some commentators on ML have suggested that, neatly corresponding with these two departments, there will be 'two AIs': one for business applications, the other for consumer devices (Economist 2017a). In Department 1 we find examples like SAP's

HANA, a ML-powered cloud database platform that enables behemoths like Walmart to monitor their entire organization's functioning in fine-grained, real-time detail (Ruth 2017), and Andrew Ng's start-up Landing.ai (founded in 2017) which aims to totally overhaul industrial manufacturing by providing 'AI-powered adaptive manufacturing, automated quality control, predictive maintenance, and more' (Landing n.d.). In Department 2, examples include various consumer commodities like Amazon Home and similar devices. Marketed as a 'smart speaker', Home is a user voice interface to the Alexa digital personal assistant that enables a variety of home automation and organizational tasks. AI is also found in other smart devices like phones and TVs and is also 'given away' as a component of free product-services such as Facebook, Twitter or YouTube where ML-based recommender systems curate timelines and give users suggestions on what to watch or listen to next. In turn, these systems gather customer data to fuel advertising revenues. However, as we will see, production of both Department 1 and Department 2 AI is often dominated by the same oligopolistic corporations, and may also be interconnected in a variety of ways, including the use of shared cloud computing facilities.

From 2015 on there has been a rapid escalation of corporate investment in AI research, venture funding of ML start-ups, and competitive hiring of AI talent as well as lots of acquisition and merger activity. Measuring the scale of this activity is difficult. According to one analysis, the AI industry had a revenue of $126 billion in 2015 and is projected to grow to $3,061 billion by 2024 (Statista 2016: 9), but another reckons worldwide spending on AI stood at only $19.1 billion in 2018, an increase of 54.2 per cent over 2017, and predicts it will reach a mere $52.2 billion by 2021 (International Data Corporation 2018). *The Economist* (2017a) calculates that in 2017 companies globally spent around $21.3 billion in mergers and acquisitions related to AI – 26 times more than in 2015. While such conflicting estimates (often manifestly driven by the self-interest of AI vendors and business consultancies) are confusing, it is clear that AI has seized the imagination of advanced capital's representatives (see also Press 2018). As *The Economist* (2017a) puts it, 'Fueled by rivalry, high hopes and hype, the AI boom can feel like the first California gold rush.'

Corporate competition for ML experts is ferocious. One study, based on LinkedIn profile data, puts the number of PhD educated people 'capable of working in AI research and applications' at 22,000, with only 3,074 currently looking for work (Gagné 2018). Demand far exceeds supply (Economist 2017b). US information capitalists are in competition both with new contenders – such as major auto companies with autonomous vehicle projects – but also now with China's tech companies, some of which have set up subsidiaries in Silicon Valley. As hiring top talent is seen as crucial for the success of AI-capital, this competition has 'set off a trend of firms plundering academic departments to hire professors and graduate students before they finish their degrees' and created an atmosphere in which job fairs resemble frantic 'Thanksgiving Black Friday sales at Walmart' (Economist 2017b). This competition for ML talent also means that wages are high.

A recent *New York Times* article reports that 'Typical A.I. specialists, including both Ph.Ds fresh out of school and people with less education and just a few years of experience, can be paid from $300,000 to $500,000 a year or more in salary and company stock, according to nine people who work for major tech companies or have entertained job offers from them' (Metz 2017b). When Google acquired DeepMind in 2014, it paid $650 million for a company of 50 employees; in 2016, the lab's 'staff costs' alone, as it expanded to 400 employees, totalled $138 million, an average of $345,000 an employee. In the light of such figures, it has been suggested that ML experts are 'the new investment bankers' (Shead 2017). The rewards are even higher, of course, for executives with experience managing AI projects. In a court case against Uber over ownership of autonomous car technologies, Google revealed that one of the leaders of its self-driving car division took home over $120 million in incentives before jumping ship to join their competitor. However, even fresh graduates with skills in ML may make 'in excess of £100,000 and sometimes up to £1 million' while still in their mid-twenties (Shead 2017).

The AI industry is international in scope. Between 2016 and 2018 it became widely recognized as a critical axis of technological competition between the United States and China, particularly given its potential for military application in an era of growing tensions. Important Chinese AI developers include its largest search engine corporation, Baidu, and ecommerce giant Alibaba (K-F. Lee 2018). Other important national sites for the AI industry include Canada, Israel and the United Kingdom. However, nearly all assessments suggest that the United States is the leading location (Jang 2017; Rapp and O'Keefe 2018; Fabian 2018). By one estimate, which surveyed over 3,000 companies around the world involved in aspects of AI development, 40 per cent are in the

US (Fabian 2018). Six, however, are preeminent: Alphabet (Google's parent company), Amazon, Apple, Facebook, IBM and Microsoft. These companies all exemplify what Tarleton Gillespie (2010) and Nick Srnicek (2016) respectively describe as 'platforms' or 'platform capitalism', a key feature of which is the digital gathering of big data generated by customers, be they users of search-engines, social media networks and video or music streaming services, or computer software or retail consumers. Access to such troves of data makes platform firms favourable sites for training ML systems.

IBM

Amongst these, IBM is in many ways an outlier, even though 'Big Blue' has a long record of interest in AI, stretching from its researchers' involvement in the famous 1956 Dartmouth workshop to the triumph of its chess-playing Deep Blue over world champion Garry Kasparov in 1997 and its AI Watson, whose 2011 victory over human competitors in the television quiz show *Jeopardy* made it briefly the public face of the new generation of AI. Yet despite IBM's $15 million investment in the system, Watson has subsequently had only limited commercial success. While it has been described as 'one of the most complete cognitive platforms available' (Kisner, Wishnow and Ivannikov 2017: 1), and has been applied eclectically to commercial ventures in fields from fashion to telecommunications, IBM's major emphasis was on potential uses in the highly profitable medical and health insurance sectors. In 2018, however, the company laid off many of the staff in this key division and announced it would be seeking new areas of focus. It is uncertain how far this setback was the result of technological failures and how much was due to the rigidities of IBM's organizational practices (Strickland 2018). IBM is likely hampered in its AI efforts due to not possessing the large proprietary pools of big data necessary for training ML systems; instead IBM has to acquire it, expensively, by buying up smaller firms engaged in medical research and data collection (Kisner, Wishnow and Ivannikov 2017: 19–20). The other major US AI producers, however, do not suffer from this problem.

Alphabet (Google)

Alphabet has been harvesting user data and applying it to advance their AI projects for years, first by algorithmically improving search patterns and matching them with ad placements, and then using similar methods for categorizing, filtering and recommending video content on YouTube or predicting which apps users of its Android mobile phone operating system would purchase. Alphabet's Google Brain unit is widely seen as the leading corporate ML research group. Between 2014 and 2018 Google bought up no less than 12 AI-related companies (Patrizio 2018), the most notable being DeepMind, which made the ML system AlphaGo that in 2016 scored an uncanny victory over the reigning human Go world champion, thus supplanting Watson as the poster-boy for AI. Such research connects not only with Google's algorithmic online services and its Google Home devices but also with its Waymo autonomous vehicles unit and suite of robotics-related companies it acquired in the early 2000s. The development of AI is an endeavour fervently advocated by Google's owners Sergey Brin and Larry Page as well as the transhumanist thinker Ray Kurzweil who is their 'director of engineering' (Simonite 2017); the combination of vast funds, deep expertise and ideological commitment places Google in an exceptional position in commercial AI research. Other US platform capitalists are, however, following similar paths.

Facebook and Amazon

Algorithmic analysis and prediction have been central to the success of Facebook in plotting the 'social graph' of users' interests and interrelations which drives its massive online advertising revenues. Facebook AI Research (FAIR) has four AI laboratories around the world, is active in conducting cutting-edge AI research, and has made several AI-related acquisitions, such as the company Ozlo, which builds virtual assistants (Patrizio 2018). Amazon's development of 'recommendations' for customers across its escalating retail and logistics operations relies on the algorithmic analysis of vast volumes of consumer data, now increasingly integrated with the operations of its huge and partially robotized warehousing and order fulfilment systems. Amazon has a number of specific AI-based products, including the Echo 'smart home' system; its digital personal assistant, Alexa; Lex, a business version of Alexa; Polly, which turns text into speech, and Rekognition, an image recognition service. In addition, ML permeates the suite of services in the Fulfilment by Amazon programme the retailer offers to third-party sellers.

Microsoft and Apple

Although of an older generation of IT companies than Google, Facebook and Amazon, Microsoft draws not only on decades of software development experience, but also on the records of interaction with millions of computer users for projects such as Cortana (the digital assistant bundled with its software), and more and less successful chatbot ventures, ranging from the catastrophic Tay (which machine-learned from online conversations to be a racist, sexist Nazi that had to be terminated with extreme prejudice) to the more innocuous teenager-imitating Zo. Finally, Microsoft's long-time competitor, Apple, which initially seemed the least AI-avid of the major platform capitalists, has also entered the arena in 2016–17, with the ongoing development of its digital personal assistant Siri and the creation of FaceID, a facial recognition security system for iPhone users. In 2018, it dramatically poached Google's AI chief (Patrizio 2018).[2]

Beyond the Tech Giants

Around the top-rank tech corporations are clustered many smaller start-up companies that are attempting to carve out niches in specialized branches of the AI industry, ranging from biotechnology, to farming, education, merchandising and surveillance (Zilis and Cham 2016; Patrizio 2018). This scene is ever changing. It is likely that, in a pattern familiar from previous cycles of IT development, many of these companies will flare up and burn out, with the successes likely to be acquired by AI industry giants: '115 of 120 AI companies that exited the market in 2017 did so by acquisition' (Patrizio 2018). Companies producing specialized hardware for AI systems also continue to appear in the shadows of dominant hardware firms like Nvidia and Intel. Venture capitalists invested $1.5 billion in hardware start-ups in 2017, twice as much as two years previously; new companies such as Cerebras, Graphcore and Cambricon have each attracted over $100 million in speculative funding. Such start-ups 'are racing toward one of two goals: Find a profitable niche or get acquired. Fast' (C. Metz 2018).

STATE ACTORS: AI SUPERPOWERS

Another crucial actor in AI development is the state. Marx never completed a systematic study of the capitalist state, and attempts by his followers to do so have stirred some of the most complex debates in Marxist theory.[3] However, one function Marx clearly did ascribe to the state in capitalism was the creation of certain general conditions of production (Marx 1990: 506; 1993: 530) such as infrastructures, which could then be transferred to private capital once they became profitable ventures. Such privatized industries in turn provide capitalist states with technological powers exercised in the name of national security against rivals: this is the dynamic of military-industrial complexes. This path of state-capital interactions is classically demonstrated by the development of digital technologies in the US. Computers and networks were incubated in the military-corporate-academic wing of the Pentagon before passing into general commercial use (Edwards 1996; Mazzucato 2013). The most celebrated, but by no means singular, case is the funding of the internet's creation by the US Advanced Research Project Agency (ARPA). Contrary to the libertarian self-presentation of digital capital, the creation of US high-tech industries depended on state-sponsored research, subsidization and contracts for the creation of the technologies which played a major role in America's Cold War victory over its state-socialist opponents.

Such state-capital interactions in the development of digital technologies were widely adopted beyond the US in the era of globalization following the end of the Cold War. While many theorists, both bourgeois and Marxist, declared the decline of the nation state, in actuality, capitalist globalization unfolded through the mediation of nation states whose activities characteristically involved the support and subsidization of the digital industries and infrastructures on which competitive participation in the world-market depended (Schiller 1999; Powers and Jablonski 2015). The so-called 'Washington consensus' governing globalization masked mounting antagonisms, in particular between a Cold War-victorious United States and the defeated 'post-socialist' Russia and China, which were compelled to capitulate to or compromise with US-led capitalism. In 2008, when the Wall Street crash manifestly weakened the US as imperial hegemon, these conflicts emerged more sharply in a flare up of economic and military rivalries to which contending nation's plans for AI are now integral.

US AI development was already built on the basis of state-initiated infrastructures and technologies; leaders of AI research, such as IBM and later Google (Nesbitt 2017), have been supported by US defence-related funding: according to one report no less than 16 US government agencies fund AI development (Fabian 2018).[4] However, it was only in 2016 that the US Federal Government began formulating an overall National Artificial Intelligence R&D Strategic Plan. The Trump administration has named AI a national priority because of 'its role in helping the U.S. lead in technological innovation as well as its role in information statecraft, weaponization, and surveillance', while the President has made it clear that he will not 'stand in the way' of AI-capital by burdening it with regulations about the social or employment effects of AI (Future of Life Institute 2018). Amongst a variety of measures, the US Department of Defense's announcement that it would invest up to $2 billion over five years towards the advancement of AI was prominent: use of ML in surveillance and for the control of 'swarming' drones and other semi-autonomous weapons appear to be a high priority (Scharre 2018). High-tech workers at Google and some other Silicon Valley companies have revolted against participation in military contracts – an important instance of resistance to AI we discuss in Chapter 2. Despite these protests, however, the Pentagon made clear that this was only the first phase of an 'AI surge' (Seligman 2018) that would include other projects such as the $10 billion to be spent over the next ten years for a special military cloud computing Joint Enterprise Defense Infrastructure (JEDI).

This sudden urgency about AI on the part of the US state is attributable to the emergence of a serious digital rival: China. In 2017, the People's Republic announced its Next Generation Artificial Intelligence Development Plan, with initiatives and goals for R&D, industrialization, talent development, education and skills acquisition, standards, regulation, ethics and security. It envisages China becoming the 'primary' centre for AI innovation by 2030 (Dutton 2018). Given the immense gulf in technological development that separated China and the US even in 2000, this ambition seems staggering. What gives it some plausibility is both the pace of China's economic growth since 2000 and changes in the nature of AI development. Kai-Fu Lee (2018), former head of Google China and now a champion of China's AI programme, argues that the most recent generation of AI – ML in particular – has passed from the moment of watershed breakthroughs to one of innovative application. Lee holds that in this scenario, China has an AI advantage in so far as it possesses large numbers of software engineers (not necessarily of superstar quality, but highly proficient), a fiercely competitive digital capital sector, and huge quantities of data gathered virtually without restraint. He suggests that the United States and China will, by 2030, constitute a state level 'duopoly' in terms of control of the AI industry. While his account ends with benign hopes for cooperation, the underlying vectors of intense economic and military competition are all too visible in his account of a 'new world order' dominated by 'AI Superpowers' (K-F. Lee 2018).

Other states are understandably reluctant to acquiesce to this vision of an America-China AI duopoly. In 2018, the European Union acknowledged that fierce international competition demanded coordinated action for it to remain at the forefront of AI development and announced a joint 'public-private' programme aimed to increase investments in AI research and development by at least €20 billion by 2020 (Middleton 2018). A striking demonstration of the centrality of state policy for AI development, and a major reason for European corporations' anxieties on this score, is the potential impact of the EU's recent General Data Protection Regulation (GDPR). This legislation places limits on corporate data gathering that are stringent by comparison with the policies of the US and China. The GDPR is criticized by business lobbyists for cramping potential ML projects – a disturbing harbinger of the potential for AI-capital to normalize and require regimes of high-surveillance governance. In an attempt to staunch a 'brain drain' of AI talent to super-salaried positions with US and Chinese corporations, the EU also launched a multinational European AI institute – the European Lab for Learning and Intelligent Systems (ELLIS) – with centres in a number of European countries (Rankin 2018). Beneath this common European front for AI, however, national rivalries continue to roil, with the UK, France and Germany jockeying for position as the regional AI leader (Shead 2018).

The US, China and the EU are merely the largest contenders in a worldwide rush by states to attract AI-capital. Japan and Russia are also substantially subsidizing AI development; by mid-2018 more than 20 nations, ranging from New Zealand to Poland, Kenya and Tunisia, had formulated 'artificial intelligence strategies' (Dutton 2018). Deep anxieties underlie such mobilization. As we will see, at the level of individual corporations, AI-capital may tend to favour winner-take-all concentration of ownership and the creation of monopolies. A similar dynamic may emerge at the level of international state relations. One of the more alarming features of Lee's predictions of a 'bipolar' US-China dominance of AI-capital is that 'while AI-rich countries rake in astounding profits countries that haven't crossed a certain technological threshold will find themselves slipping backward ... [into] a state of near total dependence and subservience', frantically 'trading market and data access' for the use of AI facilities beyond their control (K-F. Lee 2018: Kindle Loc. 2759). This prospect, along with the yet blunter threat of AI military power, ensures that no state wants to be excluded from developing the emerging general conditions of capitalist production – and destruction.

EDGE, CLOUD AND ECONOMIES OF SCALE

Essential to the AI industry, whether at the level of firm or nation, is the expensive hardware on which AI runs. Today this predominantly takes the form of 'the cloud' – vast, energy guzzling data centres that users can pay to access over the internet (Mosco 2014). As the availability of bandwidth and processing power have increased, the cloud has become available not only for storage, database and computing functions, but also in the past few years for cloud AI or 'AI as a service' (MSV 2018). AI-infused consumer devices and services send data to the cloud where the actual AI processing is done. The cloud also enables the insertion of ML techniques (like image and voice recognition) into websites or programs, as well as the online building of ML models. The intense computational requirements for training deep ML models mean that few small companies can afford to purchase the requisite hardware and instead buy computing time from cloud providers. Tractica (2018) reports that the AI industry has seen a 300,000 times increase in computing power requirements since 2012, making cloud AI a space of aggressive competition. Amazon Web Services and Microsoft Azure have been the top performers, with Google Cloud, IBM and Baidu Cloud attempting to gain market share. Cloud AI thus promises to make ML available to companies well beyond the inner circles of the AI industry, but already ownership of the cloud is consolidated in the hands of a 'very select few', with the tech giants being the dominant providers (Miller 2018).

The cloud is complemented by an emerging technique called edge computing. Edge computing is an approach in which some processing is done locally on devices rather than sent to the cloud. Tractica (2018) estimates that the amount of AI edge devices shipped will increase from '161.4 million units in 2018 to 2.6 billion units worldwide annually by 2025'. Edge computing offers advantages over the cloud when it comes to bandwidth, network latency issues and security. In applications such as autonomous vehicles, a bit of lag could be disastrous. In addition, keeping data on board a device offers obvious security benefits against hackers. As computational requirements and numbers of users grow, edge computing also becomes attractive to cloud AI providers in so far as it can reduce the load on their clouds (Miller 2018). No one, however, expects the edge to replace the cloud. The two work in tandem. Rather than a decentralized disruptor of the cloud, the edge is likely to become another axis of competition for the tech giants, who are already invested heavily in it.

In sum, the AI industry is generating many interconnected commercial ventures – an 'ecosystem', to use a term widely adopted by those who like to naturalize capitalist activity. AI is beginning to be generalized beyond the tech industry, propelled by wide applicability, financial incentives and technological advancements. The prospects for further generalization indicate to us the plausibility of a future where capital presupposes access to these means of cognition, or, in other words, where AI becomes part of the general conditions of production.

While there may be dramatic changes to the AI industry landscape over the course of this generalization, for now the tech giants occupy an apex position. Control of cloud computing facilities, ownership of large data sets, and the wealth to hire the best from a limited pool of AI talent are some of their many advantages. These factors suggest that the domination of the information technology sector by a handful of corporate behemoths will continue in the age of AI. One likely trajectory for the AI industry is thus that which Marx described as inherent to the general law of capitalist accumulation, namely the centralization and concentration of capitalist power (1990: 777). This is acknowledged even by unequivocally pro-market observers. As *The Economist* (2017a) argues: 'It seems likely that the incumbent tech groups will capture many of AI's gains, given their wealth of data, computing power, smart algorithms and human talent, not to mention a head start on investing.' And a business-oriented study enthusiastic about the commercial prospects of ML unabashedly acknowledges that there may be a 'tradeoff between innovation and competition':

> Like most software-related technologies, AI has scale economies. Furthermore, AI tools are often characterized by some degree of increasing returns: better prediction accuracy leads to more users, more users generate more data, and more data leads to better prediction accuracy. Businesses have greater incentives to build [AI] if they have more control, but along with scale economies, this may

> lead to monopolization. Faster innovation may benefit society from a short-term perspective but may not be optimal in the longer-term. (Agrawal, Gans and Goldfarb 2018: 23)

AI BUBBLE?

The latter half of the 2010s saw increasingly widespread speculation on the emergence of an 'AI bubble' (Press 2018). In a 2017 report, the business consultancy Gartner gave warning: 'As AI accelerates up the Hype Cycle, ... most vendors are focused on the goal of simply building and marketing an AI-based product rather than first identifying needs, potential uses and the business value to customers' (Hernandez 2017). It cautioned its business readers not to fall for 'AI washing' or exaggerated claims about the capabilities of AI systems. More luridly, a 2018 business report predicted that while AI would eventually 'mature' as a commercial sector, 'the field will be littered with corpses on the way' (Riot Research 2018). Might the AI industry descend into another 'AI winter' before the technology is sufficiently generalized so as to constitute a part of the general conditions of production?

The bursting of an 'AI bubble' and the failure of many of the recent entrants into the AI industry would not necessarily spell the end of AI-capital. The collapse of overvalued start-ups, and even large, established firms, making room for fresh entrants learning from their predecessors' failures, has for centuries been a normal cyclical feature of capital's innovations, from the railway and telegraph onward. This is all part and parcel of capital accumulation. The most germane recent example is the bursting of the US 'dot.com' bubble in 2000, which laid waste to many early e-commerce attempts and, by knock-on effect, threw the telecommunications industry into crisis. For a few years this disaster stalled investment in new digital business, until the emergence of Google, Facebook and other Web 2.0 companies, who, with revised business models, joined the big survivors of the bust, such as Amazon and Microsoft, in a new and even larger wave of digital commodification. A similar crisis could be a mere bump in the road along which AI advances.

There are, however, other possibilities that might more deeply disturb booming forecasts for the AI industry. One would be the discovery of serious, endemic problems in the technology itself. Roman V. Yampolskiy (n.d.), a researcher associated with the AI Safety movement, has compiled a short inventory of failures in AI. Perhaps the most serious of these dates from 2013: 'Object recognition neural networks saw phantom objects in particular noise images.' This refers to a well-documented tendency of ML object recognition systems to confuse, but report with very high confidence, certain abstract patterns with real-life entities such as peacocks or leopards. This bizarre quirk appears to derive from such systems' interpretation of radically new objects as merely extreme examples of the data sets on which they have been trained. The most immediate concern arising from this is that image-recognition systems used for military and security purposes are susceptible to malign spoofing by 'adversarial images' (Nguyen, Yosinski and Clune 2015). Even more troubling is the implication that neural networks are opaque to their producers, and can suddenly throw up entirely unanticipated results (Scharre 2018). Such disquiet has yet to seriously dampen AI enthusiasm. But disastrous AI errors might dent business confidence. A series of fatal autonomous vehicle accidents in 2018 cast a pall, however temporary, over self-driving car research by reminding corporations and publics of the complex and costly liabilities such technology involves.

There is also the possibility that AI may not deliver the goods that capital expects. The entire digital 'information revolution' has been bedevilled by the 'productivity paradox', a problem summed up in the sardonic observation by Robert Solow (1987) that 'You can see the computer age everywhere but in the productivity statistics.' Apart from a brief period at the end of the 1990s, widespread adoption of digital technologies in advanced capitalist economies has not (as of 2019) generated observable year over year productivity increases comparable with those yielded by earlier cycles of innovation. The reasons are hotly disputed. Some economists argue that this apparent anomaly reflects problems in measuring the real significance of digital activity, or that economic gains of digitization merely need more time to manifest (Brynjolfsson, Rock and Syverson 2017). Others, such as Robert Gordon (2016), insist that the economic consequences of information technology are simply much less than those of inventions such as the automobile, electricity, urban sanitation chemicals and pharmaceuticals that powered US economic growth from 1870 to 1970.

In the uneasy aftermath of the Wall Street crash of 2008, US corporate spending on new means of production, such as equipment and buildings, has been low, especially relative to the amount spent on financial and speculative activities such as dividends, stock buybacks and takeovers

(Henwood 2018a). Michael Roberts (2018) proposes that '[p]roductivity growth in all the major capitalist economies has slowed because of the failure of capitalists in most economies to step up investment in new technologies'. This picture could be changed, either by dramatic falls in the cost of AI-related technologies, by massive state subsidization of AI infrastructures, or by the threat of wage demands emerging from gradually tightening post-recession labour markets, which would incentivize automation. But there is as yet no absolute guarantee that the current wave of new AI research will actually make it out of the laboratory and into a wide transformation of business practices: the AI revolution might subside with a digitally voiced whimper.

THE GENERAL CONDITIONS OF PRODUCTION

Let us suppose that capital's current love affair with AI is not broken up by performance failures and commitment nerves. What might be the eventual offspring of this union? To elaborate the possible long-term significance of the second era of the AI industry we turn to an often-overlooked Marxian concept: the general conditions of production. What does Marx mean by this category that is mentioned in passing in *Capital* (1990: 472–3, 474, 505–6, 579, 652), but discussed in some depth in *Grundrisse* (1993: 308, 524–33, 725)?

To begin, it is helpful to understand what it means that these conditions are general. In Marx's system, 'general' is almost always opposed to 'particular', and so it is with the general conditions of production, which he distinguished from the 'particular conditions of production for one capitalist or another' that are bought or produced directly by individual capitals to keep production going, and include material inputs (raw materials and intermediate goods), means of production, and labour (Marx 1993: 531).[5] Whereas particular conditions concern the production of this or that individual capital, the general conditions of production are common to all capitals.

In *Grundrisse*, Marx spelled out the relationship between an individual capital and the general conditions as 'a specific relation of capital to the communal, general conditions of social production, as distinct from the conditions of a particular capital and its particular production process' (Marx 1993: 533). The general conditions are, therefore, 'something that benefits (or impedes) all particular capitalist production processes' (Kjøsen 2016: 65). Infrastructure is illustrative of the nature of the general conditions, because roads, canals or railways 'benefit not just a single capital, but all individual capitals in a given area' (Kjøsen 2016: 65). With the general conditions of production, Marx thus described the general milieu in which an individual capital finds itself at a particular historical moment; it is the terrain of both class struggle and capitalist competition.

Importantly, these conditions are general because they are *potentially* available to all individual capitals; this does not mean they are free or practically obtainable by all. The case is different for different conditions. Something might be a commodity as well as a general condition. For example, transportation and communication – e.g. container shipping or the hardware required to connect to the internet – are something that all individual capitals have to pay for and are necessary for contemporary capitalist production. As long as a mode of transportation is common carriage, it counts as a general condition of production. Other conditions are free or are paid for indirectly, through taxation. For instance, any and all capitals find themselves on the world market whether they like it or not. On the other hand, the protection of shipping lanes and the maintenance of transportation infrastructure are paid for by taxes.

Infrastructure is but one vital component to the general conditions of production. As one of us has pointed out, the general conditions include a bewildering array of things: the means of communication and transport; the general use of buildings for production and storage; the market, i.e. the sphere and process of circulation; the political world order; the general state of science, technology and engineering; and, confusingly, also specific kinds of production – such as the production of machinery by machinery and the degree of automation in production – as well as the mass and velocity at which production occurs (Kjøsen 2016: 64).

What is the specific function of the general conditions of production? Marx argued, while discussing infrastructure, that 'All general conditions of production ... facilitate circulation or ... make it possible ... or ... increase the force of production' (1993: 530–1). A well-designed system of highways, bridges and tunnels will, of course, facilitate circulation in the sense of making it faster or, if built where there previously were no roads, would make circulation there possible. Speeding circulation makes it possible for capital to accelerate the cycle of extracting surplus-value in production and realizing it in the market (Marx 1992: 203). Similar considerations apply to non-infrastructural general conditions of production; for example, increasing the degree of automation in production intensifies surplus-value extraction because machinery allows the application of physical force far exceeding that possessed by human labour alone (Marx 1990: 509).

To really appreciate Marx's argument about the function of the general conditions of production and thus how AI could function therein, it is necessary to discuss how they develop in lock-step with the mode of production itself. That is, the general conditions always refer to a specific locale and time period, meaning that the general conditions for the period of manufacture were different from the period of large-scale industry, which in turn differed from that of Fordism and so on, all the way up to a possible future AI-capitalism (Marx 1990: 505–6; Kjøsen 2016: 65–6). Importantly, the general conditions for one period may be 'inadequate' or 'unbearable fetters' to the following one, and will thus have to be adapted or updated so that they become 'appropriate' to the new period; when they are appropriate, 'the mode of production [i.e. period] acquires an elasticity, a capacity for sudden extension by leaps and bounds' (Marx 1990: 579; Kjøsen 2016: 65–6). That is, the velocity and volume of production can now occur at higher levels than during the previous period.

How are the general conditions adapted to emergent modes of production? To sketch the possible capitalist future of AI we need to elaborate this dynamic. The connection between the general conditions and the mode of production starts at the level of branches of industrial capital (Kjøsen 2016: 66). A revolution in production, for example through the invention of a machine, may force transformations in other branches that 'are connected together by being separate phases of a process, and yet isolated by the social division of labour, in such a way that each of them produces an independent commodity' (Marx 1990: 505). Any changes that cause productivity increases in terms of volume and/or speed in one branch require connected branches to adapt in order for the original branch to maintain its new level of productivity. To illustrate this dynamic, Marx referred to the mechanical revolution in cotton-spinning that 'called forth the invention of the gin' because without this technology, the supply of cotton would not be able to keep up with mechanized cotton-spinning (1990: 505). With the general conditions, this dynamic is writ large; paying particular attention to the means of communication and transport, Marx argues that it was specifically the generalization of production with machinery and its resultant increase in speed and output that forced a change in the general conditions to become appropriate to large-scale industrial production. Thus, when particular branches of production are closely connected, 'a revolution in terms of knowledge, technology and organization in one branch propagates throughout related branches, leading not only to growth in productivity, but also increased output, which in turn leads to new chain reactions throughout related branches of production and eventually to a revolution in the mode of production', which thus becomes elastic in its productive capacity (Kjøsen 2016: 66–7).

In addition to large-scale industry requiring an 'immense transformation' in transportation and communications networks, it also required that machines could produce large quantities of uniform, precisely tooled machine parts: 'Large-scale industry therefore had to take over the machine itself, its own characteristic instrument of production, and to produce machines by means of machines. It was not till it did this that it could create for itself an adequate technical foundation and stand on its own feet' (Marx 1990: 506). Only when machines started producing parts for machines could machinery as such become a general condition of production, meaning it was available to all individual capitals, in adequate quantity and quality. This is not to say that every individual capital *must* adopt a particular new technology for it to be considered part of the general conditions; it is sufficient that individual capitals have *access* to it on a more or less equal footing. Competition will compel adoption; it will become part of the general conditions once it is widely used by a critical mass of individual capitals. This may be encouraged or even enabled by states, as for example in the 1990s when the US government fostered business adoption of the internet as an 'Information Highway' for commodity circulation (Schiller 1999). The high level of governmental interest in boosting national AI capacities suggests that states are already pushing for another revolution in the general conditions of production. Yet before we can discuss what this push might be a response to, or what form it might take, we need to outline the general conditions of production as they stand today.

THE GENERAL CONDITIONS OF PRODUCTION FOR TWENTY-FIRST-CENTURY CAPITALISM

Marx sketched out how the general conditions of production for the period of manufacture were transformed into those adequate for large-scale industrial production. Since Marx's time, the capitalist mode of production has gone through at least two other notable periods: Fordism, defined by Taylorism and the assembly line, and the period that followed it, defined primarily by ICTs and logistics, which is most commonly described as post-Fordism (Hardt and Negri 2001). The lack of consensus around how to define post-Fordism is evident in the panoply of names that have been given to the same period: digital capitalism (Schiller 1999), logistical or supply-chain capitalism (Toscano 2011; Cowen 2014; Kjøsen 2016), and cognitive capitalism (Moulier-Boutang 2011). We recognize the amorphous quality of the term post-Fordism, and agree with many critiques made of it (see Amin 1994 for an early overview); and we particularly note that many theorists of post-Fordism have underplayed the continuities and overplayed the qualitative differences that this period has with Fordism. That being said, we also hold that it is evident that substantial changes have occurred since Fordism, such that the designation of a new period is worthwhile. Further, we suggest that we are perhaps entering a new period of the capitalist mode of production, beyond post-Fordism, which we refer to as actually-existing AI-capitalism. This, we suggest, may be seen as a middle phase of a larger mode of cybernetic capitalism (Robins and Webster 1988; Peters, Britiz and Bulut 2009; Tiqqun 2001) which tends towards fully developed AI-capitalism.

The accompanying table lays out a schematic history of capitalist modes of production and their attendant general conditions of production. Like any chart which purports to grasp complex phenomena, it oversimplifies and runs the risk of appearing more rigid that it is intended to. The periods listed here overlap with one another and were and still are being passed through in different ways and at different speeds in different places around the world.

Following Marx's theorization, we propose that a revolution in one branch of the economy could provoke the necessity of widespread AI adoption such that it becomes part of the general conditions. In Chapter 2, we suggest that a large part of the excitement about AI development responds to crises encountered by capital's globalizing search for cheap labour, and in Chapter 3 we argue that the prospect that AI may overcome the fetters to an advanced form of capital left over from a previous period – the persistent fetter of human labour – needs to be taken seriously. But here we ask how generally available AI might be incorporated into capital's inherently revolutionary dynamics. It is not difficult to understand why capital must rely, and how it relies,

The General Conditions of Production in Historical Perspective

Time Period	17th C–late 18th C	Late 18th C–mid 19th C	19th C	Late 19th C–late 20th C	1970s–2010s	2010s–?	??
Epoch of Production	Mercantile Capital	Manufacture	Industrial Capitalism		Cybernetic Capitalism		
Period of Production		Manufacture	Large-scale industry	Fordism	Post-Fordism	Actually-existing AI-capitalism	AI-capitalism
Production Technologies and Organization	Handicrafts, cooperation, hand tools	Division of labour, hand tools	Industrial machinery, division of labour	Taylorism, assembly line, mass production	Flexible production (mass customization), supply chain	Narrow AI	
Type of Subsumption[1]	Formal Subsumption		Real Subsumption		'Hyper-Subsumption'[2]		
General Conditions of Production	Canals, sail shipping, colonial markets, roads, draft animals, stage coaches	Division of labour, asphalt roads, canals, sail shipping, colonial markets and system	Steam power, machine tooling, production of machines by machines, railways, prime movers, river steamers, steamships, world market, telegraph, imperialism	Electrical power, telegraph, radio, television, automobile, Bretton Woods (GATT, WB, IMF), world market	Global and regional trade agreements (NAFTA, WTO), ICTs, networks, electronic financial markets, logistics, global supply chains, software, information, bar codes, scanning technology, container shipping and intermodal transportation, process mapping	The cloud, big data, platforms, sensors, smartphones and personal computers, GPS, narrow AI, broadband internet, web	AI autonomous vehicles, smart cities, digital personal assistants, increasingly advanced AI, production of AI by AI, 3D printing

Notes:

1. There are different interpretations of subsumption as either a logical or historical category. While this table adopts the latter, we do not all agree on using subsumption as a way to periodize the capitalist mode of production. The difference in interpretation of subsumption is related to how *Capital* is interpreted: (1) as a work depicting the historical development of capitalism (see e.g. Ernest Mandel 1990); or (2) as a theoretical analysis of capitalism that examines the 'essential determinants of capitalism, those elements which must remain the same regardless of all historical variations so that we may speak of "capitalism" as such' (Heinrich 2012: 31). For a discussion on the historical vs. logical interpretation of subsumption, see Endnotes (2010).

2. The concept of 'hyper-subsumption' is similar to Stiegler's concept of grammatization.

on physical infrastructures, such as roads and ports, to overcome the fetter of space. But what fetters to capital's valorization might generally available AI overcome? And how might AI become generally available?

INFRASTRUCTURAL AI

Our approach to the question of how AI might become part of the general conditions of production is informed by the 'infrastructural turn' in the humanities and political economy (Rossiter 2016; Cowen 2014; Steinhoff 2019a), as well as Marxist assessments of logistical and energy infrastructure (Toscano 2011; 2014; Bernes 2013; Kjøsen 2016). Such critical approaches seek to counteract the frequent invisibility of infrastructure by showing how it is implicated in varieties of power relations. However, the particular notion of infrastructural AI comes to us directly from the representatives of AI-capital. As we have seen, commentators such as Andrew Ng and Kevin Kelly expect AI to become ubiquitous, distributed by a network infrastructure, just as electricity and internet access are distributed today. Is there a fetter to production or circulation, a supply bottleneck or something else that could be motivating this particular situating of AI? No less than Microsoft (n.d.) identifies one for us in what it terms the 'fundamental constraint' of human cognitive limitations. Framing modernity as an information explosion which escalates with the computer, Microsoft laments: 'In the midst of this abundance of information, we're still constrained by our human capacity to absorb it.' The notion is that the shift to a data-centric mode of production is underway and that, as Marx argued about large-scale industry, the tech industry will not be able to 'stand on its own feet' until it creates for itself an 'adequate technical foundation' (1990: 506). This foundation is infrastructural AI – the means of cognition.

The slogan under which major AI producers advance their creation of a generalized AI infrastructure for advanced capital is 'the democratization of AI' (Microsoft n.d.; Gosaduaff 2017; Gent 2018). Microsoft, Google and Amazon have all announced projects with this goal: Microsoft has announced a plan to 'democratize Artificial Intelligence (AI), to take it from the ivory towers and make it accessible for all'. It lists the following four points, which deserve to be quoted in their entirety:

> We're going to harness artificial intelligence to fundamentally change how we interact with the ambient computing, the agents, in our lives.

> We're going to infuse every application that we interact with, on any device, at any point in time, with intelligence.
>
> We'll make these same intelligent capabilities that are infused in our own apps – the cognitive capabilities – available to every application developer in the world.
>
> We're building the world's most powerful AI supercomputer and making it available to anyone, via the cloud, to enable all to harness its power and tackle AI challenges, large and small. (Microsoft n.d.)

Microsoft imagines a world submerged in AI that is nothing short of techno-animistic: 'As we infuse intelligence into everything, whether it's your keyboard, your camera, or business applications, we are essentially teaching applications to see, to hear, to predict, to learn and take action.' Other tech giants have used the same terminology. Guy Ernest from Amazon describes Amazon Web Services as 'democratizing AI' by making their AI tools available for 'any team size and skill, and for every use case'. The very same phrase has been uttered by Fei Fei Li, Google's Chief Scientist of the Cloud and ML, and Michael Marin, a senior executive at IBM Internet of Things (Greene 2018; Simpson 2018). 'Democratizing' AI thus means generalizing both its deployment and the tools for creating it, making it increasingly available to end-users and allowing anyone, working in any field, even those without any AI training, to develop AI.

One of the most significant axes of the 'democratization' programme might seem to contradict the AI industry's profit orientation. The AI industry is characterized by a large and vigorous open-source community, in which tools and templates for making AI are freely distributed, projects are undertaken by cooperative online programming collectives, and products are released gratis for general use. Nearly all of the tech giants have open-sourced some of their AI-related materials (Simonite 2015; Crosby 2018). In 2015, Google released TensorFlow, a library of tools for deep learning programming, under the Apache 2.0 open-source licence, and it is now widely used. Other tech giants, seeing Google's success, followed suit. In 2017, Facebook opened several of its libraries, pre-trained models and data sets including its Caffe2 and PyTorch frameworks (Arakelyan 2017). Microsoft's CNTK, Baidu's Warp-CTC, and Amazon's DSSTNE (the AI framework that powers its product recommendation system) are now all freely available and can be used to produce industry-grade AI (Stone 2016; Finley 2016). In addition, there are many other open-source AI projects not derived from the tech giants (Harvey 2017). Almost all AI projects today rely on such open-source toolkits. This is a significant milestone, we suggest, for AI becoming part of the general conditions of production.

Do such open-source projects mean AI will follow a path beyond the control of giant corporations? To answer this question, it is useful to remember the history of so-called free and open-source (FOSS) programming. Free software advocates such as Richard Stallman called for the non-commodification of software in the 1980s, but were largely defeated by the business-friendly open-source movement, championed not only by Linus Torvald, inventor of Linux, but also by the likes of Tim O'Reilly, CEO of O'Reilly Media (Liu 2018; Halliday 2018). As 2020 approaches, 'open source' is a buzzword for the business press and major IT corporations have shifted from seeing the open-source community as dangerously subversive to viewing it as a source of robust no-cost programming, a potential recruitment ground, and a strategic site for attracting users to their platforms (Weber 2004; Söderberg 2008; Tozzi 2017). Indeed, some open-source projects are dominated by contributions from employees of companies who use those projects. In the case of Linux, 2017 saw 'well over 85 percent of all kernel development … done by developers who are being paid for their work' (Corbet and Kroah-Hartman 2018: 15).

The example of Google's Android operating system illuminates open-source's shift. Android was released in 2008 by Google on an 'open' basis to challenge Apple's domination of the smartphone market. While Apple's iOS remains exclusive to its iPhone, Android, in the hands of Samsung, has since 2017 become the globally dominant smartphone operating system. But what has Google gained from this give-away? Initially, it was a defence to avoid Google's search engine being cut out of an Apple-dominated mobile phone world. Subsequently, as Android itself rose to ascendancy, Google gained a widely-used platform. Google has now instituted 'closed source creep' in which 'an open source base [is] paired with key proprietary apps and services' (Amadeo 2018). Google has rendered more and more of the apps customers expect to find on an Android phone inaccessible not just to other operating services, but even to developers making versions of Android that Google does not control, such as the free Android clone Replicant. Although ostensibly open-source, in practice Android largely operates as an annex of Google's larger data-harvesting operations, which in turn sustain its massive advertising revenues and training of its ML systems. In 2018, European Union antitrust regulators fined Google a record $5.1 billion for abusing its power in mobile phone markets, declaring that 'Google has used Android as a vehicle to cement the dominance of its search engine' (Satariano and Nicas 2018).

Large corporate AI producers can thus not only coexist with, and indeed benefit from, open-source AI development, but can actually weaponize it against competitors (Vorhies 2016b). Google's TensorFlow can be run on competing (non-Google) clouds such as Amazon Web Services or Microsoft Azure, but this is likely due to Google being the latecomer to the cloud game. It is now vigorously attempting to take a slice of the market. It is possible that a Google dominant in AI could cajole or coerce developers onto the Google cloud, and TensorFlow could become 'the Android of AI' (Gershgorn 2015). Further, as the tech companies themselves admit, open sourcing software can lead to accelerated development and improvements beyond what a single team at a company could accomplish. As one commentator observes:

> Free software is good business for these companies, exactly because it allows more people to develop AI. Every big tech company is locked in a battle to gather as much AI talent as possible, and the more people flooding into the field the better. Plus, others make projects with the code that inspire new products, people outside the company find and fix bugs, and students are being taught on the software in undergrad and Ph.D. programs, creating a funnel for new talent that already know the company's internal tools. (Gershgorn 2018)

Open-source AI projects thus act as 'on-ramps' to the proprietorial infrastructures of large AI companies (Asay 2017). Indeed, corporate encouragement of open-source AI is now so high as to compel critics of digital capitalism to worry not only that corporations will quash open-source AI, but also that their encouragement of it will produce malign results, such as the 'deep fake' pornography built with Google's TensorFlow (Gershgorn 2018). So symbiotic have platform capitalism and open-source AI become since 2010 that both critiques may turn out to be true, with the commercially successful products of open-source developers gradually being consolidated under the aegis of large technology companies, while a shadow world of amateur dark-side AI is created on freely available corporate tools. These capital-open-source relationships typify what Paolo Virno terms 'the communism of capitalism' (2004: 110): corporations actively fostering a bottom-up, diversified and often free production of goods, and then harvesting this fecundity by commodifying its most successful fruits.

Such a strategy is consistent with the AI industry dynamics we have already reviewed. Major AI developers, themselves the direct and indirect beneficiaries of government-supported AI research, are both supplying AI capacities to other businesses and fostering the growth of large open-source communities. For capital as a whole, this means that AI-driven business analytics, managerial tools and production automation will become increasingly available, supplied from a combination of large cloud computing platforms and edge computing devices. For end-users, apps and products will increasingly integrate AI functions, and tools for making AI will be increasingly available and easy to use. Such a 'democratization' of AI will be entirely consistent with the reaping of massive profits by the major oligopolists of the AI industry – just as the production of earlier generations of general conditions of production, such as railways and telecommunications, created the fortunes for the corporate producers of such infrastructural technologies. If AI becomes generally available, it will still remain under the control of these capitalist providers.

THE SMART CITY, THE INTERNET OF THINGS, AMBIENT INTELLIGENCE

'Democratization' programmes are not the only way AI might be made generally available. Three other topics favoured by high-tech capital illuminate this possibility from different angles: the Internet of Things (IoT), the smart city and ambient intelligence. The IoT can be simply defined as the 'pervasive deployment of [networked] smart objects' (Kopetz 2011: 307). While the internet is commonly understood as a technology for communication between humans, the IoT is a hypothetical or emerging internet comprised of machine-to-machine communications, empowered by technologies such as radio-frequency identification (RFID).[6] The term IoT was coined by Kevin Ashton (2009), who believes '[w]e need to empower computers with their own means of gathering information, so they can see, hear and smell the world for themselves, in all its random glory. RFID and sensor technology enable computers to observe, identify and understand the world – without the limitations of human-entered data.' The goal of the IoT is for machines to dispense with human intermediaries and to communicate and act intelligently in the world on their own. Recapitulating Microsoft's thesis about the fundamental constraint posed by human cognitive limitations, some analysts suggest that such a machinic population will require AI to be functionally realized: 'the flood of data that comes from IoT devices … [has] limited value without AI technologies that are capable of finding valuable insights in the data' (International Data Corporation 2016).

The smart city is a term lacking a consensus definition (Cocchia 2014), but visions of it provide another way of thinking about AI's capitalist near future, focused on issues of urban development. Most conceptions of the smart city agree on at least one criterion: the presence of 'pervasive and ubiquitous computing and digitally instrumented devices built into the very fabric of urban environments' (Kitchin 2014: 1). These may include sensors and cameras hooked up to various actuators and processors which, through machine-to-machine communication, automatically optimize traffic, maintenance, energy distribution or various other urban flows in such ways that social life is improved. The smart city is thus a particularly urban manifestation of the IoT, but with a tacit political agenda of placing urban development increasingly in the hands of large AI-capitalists. As two critics put it, smart city discourse evinces a 'free-floating utopianism about governance as a machine that would go of itself' (Sadowski and Pasquale 2015). Google is, perhaps unsurprisingly, involved in smart city endeavours; as we discuss in Chapter 2, its project to establish a smart neighbourhood on Toronto's waterfront was met with local resistance throughout 2018, although its fate remains uncertain.

The so-called 'democratization of AI', the smart city and the IoT are all different ways of expressing the notion of ambient intelligence, or 'electronic environments that are sensitive and responsive to the presence of people' (Aarts and Encarnação 2006: 1). The goal in the ambient intelligence paradigm is a situation in which 'devices operate collectively using information and intelligence that is hidden in the network connecting the devices. Lighting, sound, vision, domestic appliances, and personal health care products all cooperate seamlessly with one another to improve the total user experience through the support of natural and intuitive user interfaces' (Aarts and Encarnação 2006: 1). The goal is that rather than a user responding to their environment, user and environment are to be engaged in an ongoing bidirectional process of interaction (Aarts and Encarnação 2006: 11). There will be a 'new intelligent intermediary layer between people and systems' (Panetta 2017) and the environment will become 'proactive' in its interactions with users (Aarts and Encarnação 2006: 11).

Generalized ambient intelligence is one way AI could become part of the general conditions of production. If achieved, it would constitute a radical change to the technological milieu of capital, particularly if the AI which becomes ambient has ML capacities for perception and cognition. What would it be like if not only human knowledge and skills were transmuted into dead labour, but if dead labour gained the fundamental capacities for perceiving and cognizing that humans have historically monopolized? Perception and cognition would, like electrification, become ubiquitous and mundane properties of things in general. Without any claims to predict the future, we can look at some existing AI applications that currently function as fixed capital and imagine a situation in which their use becomes part of the general conditions of production.

One such application is the chatbot. Chatbots are 'any software application that engages in a dialog with a human using natural language' whether textual or auditory (Dale 2016: 813). As Helen Hester (2016) notes, these apps, intended to outsource to machines aspects of clerical, administrative and communications work, 'represent, in many respects, the automation of what has been traditionally deemed to be women's labour' – a point manifest in the names and voices of Alexa, Cortana, Siri – technologies that 'do gender'. Today, chatbots are predominantly textual, but Google's Duplex (under development as of October 2018) has demonstrated uncannily human-sounding verbal conversation skills in limited settings. Some observers have even objected to its all-too-human '"um" and "ah" sounds' and demanded that it announce itself as an AI (R. Metz 2018). According to one study, 36 per cent of businesses already employ chatbots and an additional 44 per cent expect to by 2020 (Oracle 2016). Chatbots can be simple rule-based programs, but to achieve more robust functionality and ease of use, modern systems usually employ ML to learn from experience and natural language processing to converse more easily. Google's Duplex, for example, is an ML system built on a recurrent neural network (Leviathan and Matias 2018). Industry analysts expect that soon chatbots 'will use AI to manage unstructured data and complex tasks' (Panetta 2016). Chatbots are often embedded in messenger programs such as Facebook Messenger and WeChat and are most often employed in customer service applications. In its first year of opening up chatbot functionality, Facebook had more than 33,000 active bots on Messenger (Vr 2016).

The developers of Google's Duplex hope that their creation will achieve the 'long-standing goal of human-computer interaction' of 'making interaction with technology via natural conversation a reality' (Leviathan and Matias 2018). This neatly sums up the purpose of the chatbot: to act as a new interface that replaces previous customer-facing business elements, such as web stores and human technical support, with automated natural language conversation. The chatbot is intended to make online transactions as intuitive and simple as in-person transactions. Consultancy firm Gartner anticipates that '[i]nteracting with chatbots won't require any particular set-up; the technology will simply understand and do as the human asks' (Panetta 2016).

Chatbots have not only found use in customer service, however. As Lebeuf, Storey and Zagalsky have shown, software developers have been creating their 'own breed' of chatbots to 'reduce collaboration friction' in the workplace (2017: 2). As the authors demonstrate, developers use chatbots to promote the functionality of their group by coordinating schedules and tasks, promoting adherence to group norms and roles, demarcating roles, responsibilities and expertise, as well as monitoring and promoting cooperation and trust (2017: 3–4). They also use chatbots to curate large information flows and share knowledge and skills (2017: 5). In so doing, they are offloading chunks of their social activities to these bots which convert various tasks and processes into dialogic form.

2018 was a 'milestone year' for chatbots in terms of both technical advances and business applications (Seth 2018). Yet, due to the domain-specific content required for each chatbot, these systems suffer from the same problem of heavy customization workload that faced GOFAI expert systems. However, research is underway on the possibility of 'bootstrapping' or automating the production of chatbots in new fields of application using ML trained on data available in that domain (Babkin et al. 2017). If such work is successful, we can expect chatbots to proliferate further and approach the ubiquity we have posited for AI as a general condition of production.

We can also consider another existing AI application which is expected to become widespread. In 2016, Amazon opened its first branch of Amazon Go, an automated convenience store, in Seattle. As of early

2019, a tenth branch is under construction. Go, billed by Amazon as 'just walk out shopping', uses 'computer vision, deep learning algorithms, and sensor fusion' to dispense with cashiers and check-out lines (Amazon 2016). Essentially, Amazon Go relies on an array of devices for machine perception, the data from which is processed and synthesized by AI. Customers sign in to the store with the Amazon app and their actions within are tracked such that whichever items they pick up are automatically noted by the store, and upon walking out, their account is charged. Despite Amazon's automation rhetoric, as of 2019, Go stores still involve human fixers and overseers, much as autonomous vehicles do (Del Ray 2017). Melville (2017) suggests that rather than seriously delving into the low-margin grocery industry, Amazon is merely using it as a venue for testing out a new method of payment, i.e. a new interface for retail transactions. Through pervasive AI-powered machine perception, Amazon Go transforms the retail transaction from a human-to-human interaction, in the case of a cashier, or a human-to-machine interaction, in the case of self-checkout, into a commodity-to-machine interaction which remains invisible to the human customer.

Extrapolating from these examples, we can begin to imagine what it would be like if capacities for machine perception and cognition were generalized throughout society and thus became a general condition for production. Like electricity, AI perception and cognition could be put to many uses. In the cases discussed above, they are employed with the obvious goal of the replacement of human labour – we turn to the topic of AI effects on employment in the next chapter. But if we approach these cases from the perspective of limited human cognitive capacities as fetter posed by Microsoft, we can see how they might be more generally applied. In the cases of AI-enabled chatbots and automated retail, AI cognition and perception are being deployed to streamline processes of social interaction by converting them into simplified, easily digestible forms administered by machines. Complex workplace social interactions among software developers and customer service transactions are thus reduced to dialogue with a chatbot which facilitates the process. In the case of Amazon Go, the retail transaction disappears altogether from human phenomenology, transformed by the integration of various data streams.

In both cases AI acts as an interface which simplifies a complex situation. This is nothing new for the capitalist mode of production. As Vincent Manzerolle and one of us (Manzerolle and Kjøsen 2015) show in the context of near-field communication (NFC) devices, capital has enthusiastically embraced technologies which simplify and speed up transactions. Excited by AI's capacities for further simplification and speed, the consulting firm Accenture (2017) has declared that AI will be the next user interface (UI), replacing the graphic user interface that is ubiquitous on our screens today. One of us has pointed out that analysts at Accenture deploy the unusual term 'curation' to describe the novelty of AI when it is considered as an interface (Steinhoff 2019a). Accenture declares that 'at the height of sophistication, AI orchestrates. It collaborates across experiences and channels, often behind the scenes, to accomplish tasks. AI not only curates and acts based on its experiences, but also learns from interactions to help suggest and complete new tasks' (Accenture 2017: 20). The vagueness of this description confirms the imagined widespread applicability of the function of curation – an intelligent, adaptive technique of cognitive automation. AI functions as an interface which not only represents some information to a user, but by perceiving and cognizing, actively gathers, processes, reveals and hides information before and while the user acts on it. This is the imagined function of the 'new intelligent layer' posited by ambient intelligence advocates (Panetta 2017) and Kelly's 'cheap, reliable, industrial-grade digital smartness' (2014).

Recall that, for Marx, infrastructure functions to 'facilitate circulation or even make it possible at all, or ... increase the force of production' (Marx 1993: 530–1), while the function of the general conditions as such is to give the mode of production an elasticity, i.e. a capacity for expanding the volume and velocity of production. Roads and digital networks facilitate circulation. Education systems increase the force of production by producing and distributing knowledge and ideology. In the context of capital's collision with the fetter of limited human information-processing capacities, infrastructural AI is being positioned to curate information flows, saving the human cognitive apparatus for whatever machines cannot yet handle. The retail transaction, essential to capital, yet annoyingly (for capital) consumptive of time, thought and wages, can be etherealized and accelerated with Amazon Go's AI. This transaction is effectively being curated out of existence by Amazon, and while human agents of Amazon directed this process, more advanced AI systems may develop and enact their own processes of curation, or these might arise from the interaction of multiple AI systems. It is not possible to enumerate here all the ways an infrastructure of AI could optimize circulation and/or increase the force of production. Already in the works are systems such as SIX's (n.d.) 'complete infrastructure for ecommerce over the Internet of Things', which aims to link smart home appliances (such as refrigerators which perceive when supplies are running low) with market information, retailers and delivery services to completely eliminate shopping from the purview of human cognition and action. There are certainly many other ways in which capital's valorization and realization could be powered-up by offloading cognitive and perceptual capacities to the environment at large.

Keeping in mind that curation is only one possible application of infrastructural AI, we suggest that AI could well become, and is being positioned by capital to become, a part of the general conditions of production. If this occurs, AI would become a cognitive analogue to the means of transport and communication – the *means of cognition*. The means of cognition would be a new layer of technological infrastructure interlaced with both the means of production and the means of transport and communication. While capitalist production has always relied on the human capacity for cognition in both conception and execution, an infrastructure of AI would allow the distribution of cognitive and perceptive tasks to machines, which would perform them in different, machinic ways, with potentially revolutionary effects on the mode of production. Just as Marx felt it necessary to specifically mention the then-new types of the means of communication and transport (steamships, railroads and the telegraph) because of their pivotal role in enabling large-scale industry, we single out the means of cognition as the factor which might come to define a new mode of cybernetic production.

In establishing the means of cognition, capital would, without metaphor, gain the ability to think and perceive. Marx noted that 'The development of the means of labour into machinery is not an accidental moment of capital, but is rather the historical reshaping of the traditional, inherited means of labour into a form adequate to capital' (1993: 694). In the possible future we have outlined here, capital is reshaping the primordial capabilities of the capacity to labour – cognition and perception – into machinic forms adequate to capital. The push to 'democratize' AI may be read as capital's effort to accelerate the capture of these fundamental capabilities and their implementation in machines distributed throughout society.

THE MEANS OF COGNITION

Capital is far from ready to jettison humanity completely, but it is prepared to imbue more and more of the cognitive capacities of humans into machines that form an increasingly 'smart' technological environment. AI is not only significant for its potential to automate work but also in so far as it could become part of the general conditions of production. If the means of cognition are established, production will certainly become increasingly automated, and it will become so within an environment where intelligent machines are perceiving, learning and communicating. As ML improves and new AI techniques appear, this network of intelligent machines will further expand its capacities, and capital, across the social factory, will become increasingly aware, intelligent and communicative.

In proposing the means of cognition, we might seem to be aligning ourselves with *post-operaismo* theories of cognitive capitalism (Vercellone 2006; Moulier-Boutang 2011; Hardt and Negri 2017).[7] On the contrary, we hold that the means of cognition scenario outlined in this chapter is incompatible with *post-operaismo* analyses, within which cognition is strictly anthropic. In fact, we argue that the means of cognition scenario entails a complete revision of the notion of general intellect which *post-operaismo* appropriated from Marx.

Marx described the general intellect as manifested in technology, which, in the capitalist mode of production, largely appears in the form of capital. Marx's major mention of the general intellect famously goes as follows: 'The development of fixed capital indicates to what degree general social knowledge has become a direct force of production, and to what degree, hence, the conditions of the process of social life itself have come under the control of the general intellect, and been transformed in accordance with it' (1993: 706). Here the knowledge of the social individual becomes a direct force of production, but *only* (and this point is often overlooked) once it has been implemented in machinery, because it is only in an object external to the human body that human skills and knowledge can be social, i.e. generalized. Today, anyone has the ability to instantly, if roughly, understand phrases in dozens of languages – if they access Google Translate via the Web or a smartphone app. The general intellect refers to this ever-increasing manifold of skill and knowledge possessed by capital in machinic form. Prior to its implementation, the skill and knowledge of the social individual, on the other hand, was called by Marx the 'social brain' (1993: 694). Capital continually excises and emulates aspects of the social brain, implementing them in machines and adding them to the general intellect. This is a planetary-scale version of the 'knowledge capture' that management theorists define as 'turn[ing] knowledge that is resident in the mind of the individual into an explicit representation available to the enterprise' (Gartner n.d.). The more of the social brain that the general intellect captures, the more powerful capital becomes.

Post-operaismo has rejected Marx's original formulation of the general intellect. Virno redefines it as 'the linguistic-cognitive faculties common to the [human] species' (2004: 42). This redefinition is necessary, he says, because:

> Marx conceives the 'general intellect' as a scientific objectified capacity, as a system of machines. Obviously, this aspect of the 'general intellect' matters, but it is not everything. We should consider the dimension where the general intellect, instead of being incarnated (or rather, *cast in iron*) into the system of machines, exists as attribute of living labor. The general intellect manifests itself today, above all, as the communication, abstraction, self-reflexion of living subjects. It seems legitimate to maintain that, according to the very logic of economic development, it is necessary that a part of the general intellect not congeal as fixed capital but unfold in communicative interaction, under the guise of epistemic paradigms, dialogical performances, linguistic games. (Virno 2004: 65)

For Virno, the general intellect is the novel social capacities of networked human beings, which include communicative interaction, abstraction, dialogical performances, linguistic games, cooperation and communicative competence (2004: 65) as well as an 'infinite variety of concepts and logical schemes' (2004: 106). We might also add affect, which fellow *post-operaismo* thinkers Hardt and Negri depict as a characteristic quality of work in the era of the general intellect (2001: 292). In post-Fordism, capital depends on these capacities of labour, which it cannot control or emulate. Thus, 'a decisive role is played by the infinite variety of concepts and logical schemes which cannot ever be set within fixed capital, being inseparable from the reiteration of a plurality of living subjects' (Virno 2004: 106). The notion is that in the era of the general intellect, human labour enjoys a renewed array of powers which capital cannot displace with machinery.

This chapter has demonstrated that a project is underway to make AI capacities for cognition and perception generally available. But could it be true that Virno's impossible-to-automate human capacities are indeed such? We do not believe so. This chapter has already surveyed some examples of those capacities being offloaded to AI systems, but we address them now directly.

Virno emphasizes the human capacities for communication. Communication (between machines and between humans and machines) has been one of the most successful applications of AI research. Natural language processing, speech synthesis and chatbots are only some of the ways in which AI enables the implementation of communicative capacities. As far back as 2012, Narrative Science was using AI to write formulaic sports and finance stories for newspapers and news websites. In 2019, the company is now selling the product Quill, which processes big data with Natural Language Generation (NLG) to create narrative-structured reports for businesses.

Related to communication is cooperation, which Virno also cites as distinctively human. AI is only beginning to implement cooperation, but already the examples are compelling. Generative adversarial networks (GANs) are a recent AI development in which two ML systems engage in an adversarial relationship which can be compared to the relationship between counterfeiters who produce fake currency and police who attempt to detect it (Goodfellow et al. 2014). One system attempts to generate data which closely resembles, but is not included in, a training set. The other system attempts to pick out fakes. Both systems learn over many instances of this exchange and eventually 'both teams ... improve their methods until the counterfeits are indistinguishable from the genuine articles' (Goodfellow et al. 2014: 1). This is a kind of rudimentary cooperation which produces novel outputs.

A more vivid example comes from OpenAI who have constructed a team of five neural networks called the OpenAI Five which work together to play the popular multiplayer computer game Dota 2 (Defence of the Ancients 2). In 2018 the five networks defeated a team of professional players (Park 2018). The developers explain that the team members cooperate not by explicit communication: instead, 'Teamwork is controlled by a hyperparameter we dubbed "team spirit". Team spirit ranges from 0 to 1, putting a weight on how much each of OpenAI

Five's heroes should care about its individual reward function versus the average of the team's reward functions. We anneal its value from 0 to 1 over training' (OpenAI 2018). Here cooperation results from the optimization of numerical values.

What about the capacity for abstraction or the ability to generate and grasp an 'infinite variety of concepts and logical schemes' (Virno 2004: 106)? The pioneers of deep learning address this directly when they describe the distinctive capacity of deep learning systems as 'learn[ing] representations of data with multiple levels of abstraction' (LeCun, Bengio and Hinton 2015: 436). Or as a tutorial from IBM puts it: 'By building multiple layers of abstraction, deep learning technology can solve complex semantic problems' (Sedlak 2016). This is not mere hype. Machinic language translation is now done with neural networks because the networks are able to find patterns and develop relevant linguistic concepts (rules) which produce better translations than any other automated technique. In addition, deep learning systems are immensely scalable, meaning that the more data that becomes available, the more the performance of these systems increases (Ng 2015). The performance of older algorithms plateaued at a certain amount of data, but, so far, deep learning has not. The epistemic, logical and semantic complexity that such systems will be able to grasp in the future is impossible to predict.

Finally, we come to affect, a capacity historically inimical to machines. While there is, as of 2019, no evidence of machines expressing affect, the ability of AI to capture and process data on the emotional states of human beings continues to improve. AI techniques are being deployed in the field of sentiment analysis, which seeks to computationally recognize subjective states including emotions and opinions in text (Pang and Lee 2004), in emotion recognition in speech (Morrison, Wang and De Silva 2007), and in recognition of facial emotions (Zhang et al. 2017). While Japanese companies are investing heavily in care robots, a wide variety of businesses are becoming interested in using AI to monitor and manipulate the emotions of their customers (Faggella 2018a). AI may not experience joy or sadness any time soon, but AI will likely be increasingly used by capitalist firms to add affective dimensions to diverse facets of production, circulation and beyond.

As these examples show, the capacities Virno reserves for a human general intellect are in the process of AI-driven implementation. What place then might human labour find within AI-capital, if not the one reserved by *post-operaismo*? A passing remark made by Marx is suggestive. In the 'Fragment on Machines', Marx imagined a highly automated workplace where 'the human being comes to relate more as watchman and regulator to the production process' (1993: 705). Here a human attendant watches over the lights-out factory, present only to press the stop button if something goes awry. The rarely quoted following sentence (which is interestingly bracketed) reads: 'What holds for machinery holds likewise for the combination of human activities and the development of human intercourse' (1993: 705). The term 'intercourse' was translated by Martin Nicolaus from the German '*verkehrs*', but another translation would be 'traffic'. As Matthew W. Bost shows, Marx used *verkehr* to refer to all kinds of intercourse including biological reproduction, linguistic and semantic exchange, commercial exchange, as well as 'the exchange of weapons fire in war' (2016: 338). Thus, here Marx seems to suggest that humans may also come to relate as mere watchmen to the development of their societies and the infrastructures (broadly understood) which underpin them. The offloading of cognitive capacities entailed in the establishment of the means of cognition could provide the means for capital to achieve such deep control.

Post-operaismo's anthropic, rather than machinic, conception of the general intellect not only inverts Marx's own formulation but is unable to account for actually-existing applications of AI. In addition, emphasizing a human general intellect leads to an overestimation of the ease with which revolutionary subjectivities, such as Hardt and Negri's (2001) 'multitude', can mobilize against AI-capital. The question of class power today requires a detailed analysis of what AI can really do, whether for or against labour. The next chapter investigates the possibilities for working-class recomposition in this light, which will lead on to a reconsideration, in our Conclusion, of the original *operaismo* strategy of 'refusal' – rather than *post-operaismo* 'reappropriation' – of capitalist technological development.

In sum, while capital remains necessarily underpinned by human cognition and perception, there is a project underway to replace these basic capacities with an infrastructure of AI. The establishment of the means of cognition may be interpreted as the project of, quite literally, infusing the logic of capital into the world, so that capital, instead of the humans situated within it, may think and perceive.

2
Automating the Social Factory

ROBOTS CAN'T BEAT US YET?

In the summer of 2018, the cable television company HBO ran season two of its sci-fi TV hit *Westworld*. Androids programmed to enact the sex-and-violence fantasies of visitors to a frontier-era theme park revolted, extracting a revenge of blood and fire while sorting out the existential dilemmas of machinic self-awareness amidst the stunning buttes, mesas and giant cacti of the US South West. As this drama unfolded on-screen, a more mundane dispute was underway in US capitalism's real-life desert entertainment centre. On 1 June, a contract negotiated by the Culinary and Bartenders Unions, representing some 38,000 bartenders, cocktail servers, maids, cooks and other staff working in Las Vegas for large hospitality companies, including MGM and Caesars, expired. Wage increases, workload quotas and sexual harassment were on the table, but a prime issue was automation. Machines had long been eroding casino jobs. 'Change-girls' had been replaced by cash dispensers on the gambling floor and cooks reduced by mass food preparation. Approaching over the horizon were robot room services, digital check-in desks and touch-screen cocktail orders. When US manufacturing jobs dwindled due to automation and offshoring, a widely cited measure of deindustrialization was that there were more people working in casinos – the quintessential 'service sector' job – than in car factories. Now the automata were coming for the casino workers. Union representatives spoke of bargaining for separation settlements or robot-tending retraining. Some workers were more confident – sort of: 'Sooner or later it's going to happen, but robots can't beat us yet', a barman resolutely declared (Hernandez 2018). The denouement of this conflict was not, however, as dramatic as *Westworld*; repeating a pattern that has brought US strike rates to historic lows, the union, despite a bravura display of picket preparation, shrunk back from confrontation.

'Will a robot take my job?' is a question that encapsulates the fears and hopes most commonly associated with machine intelligence. This chapter addresses the AI and jobs issue, but does not take for granted the assumptions about the justness or inevitability of labour market relations that usually frame such discussions. Rather, it sees issues of AI, jobs and joblessness as matters of class power inseparable from the chronic conflict of capital and its workers. We present two concepts useful for thinking about such conflict, 'class composition' and 'the social factory', and apply them to analyse the origins of capital's current AI-enthusiasm in the financial crash of 2008 and the subsequent crisis of globalization. Then we look at the class relations of the AI industry, whose workforce divisions and deepening automation has, we suggest, much to tell us about the wider work-world AI may make. We then go on to show how, while 'artificial general intelligence' of the sort required for *Westworld* androids currently remains the stuff of sci-fi, 'narrow AI' is changing or eliminating many types of work. Mainstream debate about the direction of this transformation pits futurist predictors of an 'AI Apocalypse Now', who foresee an imminent crisis of employment, against economists arguing that capitalism can continue with 'Business as Usual', disturbed by only minor tremors. Though this debate is speculative and inconclusive, AI raises many other issues – around worker monitoring, precarious labour, the polarization of incomes and work conditions, the availability of education and training, and the conditions of social interaction – that have major implications for the balance of class power and the possibilities of radical political organizing. Outbreaks of resistance to actually-existing AI-capital are appearing in many quarters, but the difficulties these face should not be underestimated.

CLASS COMPOSITION AND THE SOCIAL FACTORY

The idea of 'class composition' (Zerowork 1975; Kolinko 2002; Notes from Below 2018a, 2018b) originates in what is known as 'autonomist Marxism' (Cleaver 1979; Dyer-Witheford 1999; Eden 2012), a school whose emphasis on the potential strength and autonomy of workers (hence 'autonomist' Marxism), supplies a counterpoint to discussions of capital's deepening machinic powers. Marx held that capital had a long-term tendency to replace workers (variable capital) with machines (fixed capital), and discussed this both in 'technical' terms, as it altered the organization of work, and in 'value' terms, as it affected capital's rate of profit. For autonomist thinkers, Marx's formulation was one-sided. It lacked a theory of how workers *resist* machinic exploitation. So they inverted Marx's (1990: 762) concept of the 'composition of capital' and looked instead at the composition of the working class, that is to say, its changing capacities to challenge domination by capital.

Like Marx, autonomists looked at the 'technical composition' of production, such as the division of labour, management practices, the spatial layout and temporal rhythms of workplaces, and especially the use of machinery. They did so, however, to emphasize the problems and possibilities of 'political composition', that is, the organization of the working class to fight for improvements in wages, hours and conditions, and, ultimately, create a revolution against capital. Political composition might take the form of trade unions and socialist or communist parties; as or more important were waves of wildcat strikes, sabotage and absenteeism. Such rebellions would often be led by workers in some particularly strategic sector of the economy. Capital, on the other hand, would attempt to 'decompose' or break down such movements, replacing striking workers with machines, raising the overall unemployment level, or surveilling workplaces to repress organizing. Autonomists think capital cannot do without humans, and, given this, decomposition will always be met by recomposition as workers, often within some new sector of industry, perceive the cracks and weaknesses in capital's latest methods of control.

The sequence of composition/decomposition/recomposition thus results in 'cycles of struggle' that drive capital on an incessant flight into the future, seeking ever-more extreme solutions to its worker problem. Historically, the resistance of skilled technical workers to early industrial capital was slowly decomposed first by the time and motion studies of Taylorism and then by the semi-automated assembly lines of the Fordist factory. But this change paradoxically became the basis for a working-class recomposition. By reducing work to the homogeneity and monotony of the assembly line, Fordism generates the 'mass worker', whose power lay in the ability to bring to a halt the huge technological apparatus in which it is implanted. The mass worker's parties, unions and strike-power terrified twentieth-century capital with the threat of revolution, a threat that, even when unrealized, extracted the concession of regular wage increases and welfare-state provisions.

This brings us to a second term from the autonomist lexicon, 'the social factory'. Marx (1990, 1991, 1992, 1993) described capital as a circuit, a total system comprising both the sphere of production and that of circulation, with neither holding primacy. Whereas production is the dual process of making use-values and extracting surplus-value from labour, circulation realizes this value by getting commodities to market and selling them, a process involving activities such as transportation and logistics, advertising and retail sales. Marx also identified a third, increasingly powerful sector tied to circulation, that of finance: banking, credit and speculative activities. Finally, he at least sketched the sphere external to, yet functionally necessary for, capital, that of consumption and social reproduction in which human capacities to labour were renewed. He thought of this in terms of the consumption of goods, but, as feminist Marxists have forcefully pointed out, it also involves vast amounts of work, much of it unwaged and performed by women bearing, raising, educating and maintaining people for a life of labour (Federici 2012; Bhattacharya 2018). 'Social factory' is the term used by autonomists to name this entire complex of integrated functions. Initially, they used it to indicate how the industrial factory was a hub around which all capital's other activities revolved (Tronti 1977). Later they developed it to suggest how capital could be fought not just on the industrial shop-floor, but in schools, households, shops and warehouses around the entire circuit of capital, thereby expanding the concept of the proletariat beyond a solely industrial base.

Ironically, autonomism articulated its theories of worker power in the social factory at the very moment when capital's adoption of cybernetic technologies began to present an unprecedented challenge to their fundamental premises. Computers and digital networks had been developed within the US military-industrial-academic complex during the Second World War and the Cold War. From the 1970s on, in the midst of an economic slow-down that saw profit rates sharply decline after a 30-year boom, these technologies were increasingly deployed at the home fronts of North American and European capitalism. This transition from Fordism to post-Fordism decomposed the power of the industrial working class. Beverly Silver (2003), a labour historian who, while not herself an autonomist, has a broadly similar analysis, describes how in confronting its labour problems, capital can resort to three 'fixes': 'technical', 'spatial' and 'financial'. From the 1970s on, the 'technical fix' entailed automating factories and offices, pursuing the mechanical liquidation of labour, but at a level heightened by digital technologies, exemplified by the introduction of robots into auto factories. The 'spatial fix' involved relocating industrial production via supply chains to cheap labour/low-regulation zones of the world market: rather than replacing capital's workforce with machines, this fix expanded it globally, but at the lowest wage, and with maximum disposability. The third 'financial fix' attempted to escape production entirely by moving into the realm of speculation, developing instruments such as derivatives and futures, at first to defensively hedge foreign investments, then morphing into high-risk gambling machines.

Automation, supply chains and financialization gutted factories, stalled wage growth and extinguished industrial militancy in the global North. Combined with political measures to weaken unions by governments such as those of Reagan and Thatcher, they broke the strength of the mass worker. As once prosperous cities became industrial rust belts, employment within the zones of advanced capital increasingly moved to the service sector, a term covering manifold fragmented and divided forms of work, from finance to trucking to domestic care, some well remunerated, others pitifully paid, or at best with stagnating real wages. There was also a change in the gender composition of labour as households became increasingly dependent on women performing both paid and unpaid labour, a shift that many left organizations have even yet failed to come to terms with. An entire political culture based on the power of industrial labour, its unions and socialist parties sank, like Atlantis sliding beneath the sea.

With globalization, the social factory in some ways becomes a 'planet factory', with a hugely expanded circulatory apparatus of highways, container shipping, air transportation, warehouses and distribution centres, fibre optic cables and data centres, moving goods, people and information around the world. Proponents of globalization, and even some autonomists (Hardt and Negri 2001), speak of a 'flat' or 'smooth' planetary economic order. But in fact the capitalist planet was divided by sharp hierarchical zonings. Advanced regions, such as North America, Europe and Japan, housed business headquarters, financial activities, high-technology research and design, and retail sales and consumption complexes. Newly industrializing areas, especially China, became centres of relocated manufacturing. Petro-states, from Russia to Saudi Arabia, fuelled the carbon economy. Other regions sank into immiseration, with survival depending on informal or criminal activities. Class composition became hugely complicated in an increasingly unified world market that nevertheless divided workers whose standards of living and conditions of social reproduction remain universes apart (Woland, Blaumachen and Friends 2014). We therefore continue to refer to the 'social factory' internally constituted by regional arrangements, even while recognizing their global articulation.

Globalization and digitization changed the balance of class power in capital's favour. Around the world, labour's share of GDP relative to that of capital steadily decreased, in rich and poor countries alike (Economist 2013). Then, in 2008, the planet factory burst into flames. Capital's increasing focus on financial speculation culminated in the Wall Street subprime mortgage crash. Widespread contagion of financial systems, bank bailouts, austerity, recession in the global North and ramifying knock-on effects around the world followed. So too did social tumult. In 2011 a series of rebellions against inequality and insecurity – often in the form of public square occupations – ricocheted from Tunis and Cairo to Madrid and New York, and then to Istanbul and Rio de Janeiro, Tel Aviv and Kyiv.

While these revolts initially indicated a burgeoning working-class recomposition, often drawing on new strata of post-Fordist cultural and technological workers and students, they were almost all unsuccessful; they fizzled out, were repressed, replaced one conservative regime with another, or imploded into civil wars and foreign interventions. They were followed by another wave of turmoil, driven from the right: the UK Brexit referendum, Trump's 2016 US election victory, and the rise of neo-fascism across Europe, all propelled by resentments against capitalist globalization, racist hatred of migrants and minorities, and working-class anxieties around job losses and flagging wages.

Both the unsuccessful recompositionary moment of the Occupy movements of 2011 and the all too successful decompositionary neo-fascism of 2016 were symptoms of a weakened working class; but they also signalled an alarming instability within capitalism. The wave of corporate interest in AI, though long in the making, surfaced towards the end of this roller coaster of unrests. Machine intelligence was framed as the 'answer' to crisis. Thus consulting firm Accenture began a 2016 pitch for AI investment by acknowledging a 'marked decline' in the ability of capital to 'propel economic progress' and 'sustain the steady march of prosperity enjoyed in previous decades' before proclaiming that 'AI has the potential to double ... annual economic growth rates by 2025' (Purdy and Daugherty 2016).

Paradoxically, the digital industries that enabled globalization were strengthened by its 2008 near-death experience. With Wall Street reeling and Main Street industries, such as the big three US automakers, mired in recession, 'big tech' was the one beacon of hope. Having survived the dot.com meltdown of 2000 by reincarnating as 'Web 2.0', digital capital was already on the ascent before the Wall Street meltdown. Google first offered stock market shares in 2004 and Facebook in 2006. Apple released the iPhone in 2007. It was, however, during the post-crash recession that these digital corporations, along with others such as Amazon, topped the lists of the most highly valued companies in the world, occupying positions previously held by giant energy and financial service corporations (Mosco 2017: 65). Even uprisings against global capital, hopelessly simplified as 'Facebook revolutions', could be interpreted as manifestations of a new digital order. The US return from recession after 2008 was a 'tech-led' recovery, during which the preconditions for ML were established: big-data-generating social media, web and smartphone use.

Very specific aspects of the crisis of globalization boosted interest in AI. Amid the noise of round-the-world protests, one clear message could be detected: the days of cheap global labour were fading. While revolts in the global North West failed, something else was happening in the South. New generations of industrial labour were finding their own political composition. In China, successive strike waves in the factories of the Pearl River Delta steadily drove wages up. In 2010 protest-suicides in the giant electronic plants of Foxconn, where Apple's iPhones were assembled by a workforce under quasi-military discipline, flashed around the world. Foxconn's exasperated owner, Terry Gui, threatened to replace his assembly workers with one million robots. With the first influx of migrants from rural villages to cities exhausted, China's factories found it increasingly difficult to attract young people to assembly line drudgery. Average hourly wages for China's manufacturing workers trebled between 2005 and 2016, approaching the hourly rates paid in countries like Greece and Portugal (Gao 2017). This was only one sign of wider problems for globalized capital: in India, striking auto workers burned factories; in Cambodia, garment workers confronted police in the streets. With the sudden implosion of the financial fix, and the slow erosion of the spatial fix of worldwide low-wage locations, capital started to look at a new, more intense technical fix: AI automation.

Other, yet more violent crisis-events drove AI forward. The '9/11' destruction of New York's World Trade Centre in 2001 and the subsequent 'war on terror' had stimulated the development of ML technologies to manage vast surveillance systems. Wars in Afghanistan and Iraq continued this momentum, adding drones and other autonomous vehicles using rudimentary machine intelligence to the US counterinsurgency arsenal. Military interest escalated when popular uprisings in Ukraine, Syria, Libya and elsewhere collapsed into civil wars and foreign interventions. As the US, Russia, China, Israel, Saudi Arabia and Iran squared off in shadow-wars, research and application of ML for surveillance, hacking and cyber-security also accelerated (Sanger 2018; Scharre 2018: Dyer-Witheford and Matviyenko 2019). Just as the Cold War had incubated digital technologies that then found their way into civilian production, so too did New Cold War weaponry work its way back into the front lines of capital's confrontation with labour, as drones and improved robots entered the workplace.

As significant to the AI industry was the widespread perception that US and European capitalist liberal-democracy was in trouble. Precarious workers occupying Wall Street and rust belts voting for Trump or Brexit brought home the point that the advanced capitalist economies were not delivering growth sufficient to both give corporations big profits and raise working-class living standards. Despite utopian promises about computers and networks, digital technologies had not been able to repeat this amazing double act of Fordism. Silicon Valley's response was to double-down on its cybernetic bet. ML and advanced robotics would be magic bullets for renewed economic growth (Purdy and Daugherty 2017; Swanson and Mandel 2017). Such promises are appealing. What they obscure is that for AI investments to be undertaken by capital, they have to ensure not just economic growth, but profitable growth, a surplus well over anything accruing to labour. And this might require making labour more, not less disposable, and less, not more capable of claiming improved wages and working conditions and of exercising any type of autonomy vis-à-vis capital. It could, in short, entail further decomposition of the working class, which arguably never recomposed after the 1970s.

THE LABOUR OF AI

It seems paradoxical to speak of the labour of AI, a technology that purportedly eliminates human work. Yet AI itself has to be produced before it is applied in other industries. This is both a precondition for the AI-driven transformation of other types of labour, and provides some clues as to the direction that a more general recomposition of work may take. There are many different kinds of AI research currently in play, and while there are broad similarities in their production processes (e.g.

dependence on highly trained software engineers and programmers) each has its particularities. Here we focus on ML, the leading form of AI in the second decade of the twenty-first century.

Catherine Dong (2017) explains that 'Machine learning engineering happens in three stages: data processing, model building and deployment and monitoring.' The 'meat' in this 'sandwich', she says, is the building of the model. This is where software engineers develop the algorithms that recognize given input data, be it digital images of cats, or of pedestrians crossing streets, or configurations of search terms, or purchase records, or social media topics. This learning process is tested, tweaked and improved over thousands or millions of training examples, and then further tested on raw data, where the pattern signal is mixed with noise (images of cats with those of dogs, antelopes and elephants; pedestrians alongside wheelbarrows, letter boxes and unicyclists, and so on). At the end of this process the AI should be capable of identifying its target object with an acceptable level of confidence and of establishing statistical correlations between different patterns (more pedestrians crossing streets in daytime than at night; cat images posted more frequently by women than men, etc.). It is these capacities that give ML its uncanny, inhuman ability to predict associations and outcomes.

There are many techniques to construct algorithmic models – 'linear/logistic regression, random forests and boosted decision trees' – familiar to computer scientists for some time (Dong 2017). These are used to create the friend-suggestions, ad-targeting, recommendations and search rankings that are part of everyday internet use. However, the most sophisticated ML modelling involves 'deep neural networks' that are 'hard to train ... require[ing] more time and computational power' and 'a combination of intuition and trial and error' (Dong 2017). While courses and do-it-yourself instructions in ML are proliferating, and libraries of open-source tools widely available, deep expertise with neural networks is rare; capable computer scientists may be numbered only in the thousands worldwide (Economist 2017b). This accounts for the intense corporate bidding wars for such talent and the high salaries and benefits it can command. ML thus intensifies the basic software development labour process in which core activities are conducted by small groups of highly skilled techno-scientific workers.

While the 'meat' of ML is the making of the algorithmic model, it is 'the bread of the sandwich process' – what happens before and after training the algorithm – 'that holds everything together' (Dong 2017). These two layers of the ML process involve, at one end, 'cleaning and formatting vast amounts of data to be fed into the model' and, at the other, 'careful deployment and monitoring of the model'. As Dong observes: 'Any incompatibility from any stage of the ML development process – from data processing to training to deployment to production infrastructure – can introduce error' (2017). In practice, ML software engineers often spend *more* time dealing with 'preparing and monitoring' than in actual model making. Later in this chapter we will address the sources of such data sets, from unknowing mobile phone users to closely monitored distribution centre workers and the sleepless 'safety drivers' in driverless vehicles. But regardless of how data is collected, it nearly always has to be cleaned for use in AI training: standardizing its format, eliminating errors, detecting outliers, supplying additional information, adding features used in processing and above all labelling objects that ML will learn to recognize. Amongst data scientists, jokes such as '80% of the job is cleaning data and the other 20% is complaining about how the data has been cleaned' abound (Lohr 2014).

Much of this 'data munging' or 'data wrangling' (Johnston 2015) is outsourced, often as offshored microwork, in low-wage zones from Venezuela to India and Indonesia, at rates ranging from a few dollars to a few cents an hour (Nakashima 2018a). At first this was handled through crowdsourcing platforms such as Freelancer, Mechanical Turk or Upwork. Then data cleaning became a specialized business. Start-ups like Mighty AI and Figure Eight (formerly CrowdFlower) develop software that 'makes it easier to label photos and other data, even on smartphones' (Nakashima 2018a). Playment, a Bangalore company, 'gamifies' such tasks and calls its 30,000 on-call workers 'players'. In China, firms specializing in 'refined data processing' work with a Taylorist division of labour. Some 'project groups' are responsible for labelling 'nodes on the human body in complex yoga-like postures', or annotating 'motorcycles, bicycles and pedestrians' showing their travel directions for autonomous vehicle software; others 'scan the contours of buildings or obscured objects, used to train radar to detect real-world objects' (Yu 2017). The work is demanding and tiring.

The other layer of the ML 'sandwich', deployment and monitoring, also draws on global clickwork. ML algorithms may not work as anticipated, or encounter problems requiring intervention. How much purportedly automated systems depend on human backstops was highlighted in 2017, when Expensify, a company offering to automatically scan photos of receipts to extract data for expense reports was discovered to have bridged failures in its text analysis software by sending customers' personally identifiable receipts to workers on Amazon's Mechanical Turk (MTurk) crowdsourcing platform. This kind of 'human safety net' behind AI-powered services is very common (Gent 2017). Sarah Roberts (2016, 2017) has identified the role of generally poorly paid and precarious human content moderators in filtering social media feeds to purge violent and pornographic content. Google is reported to have an army of 10,000 'raters' watching YouTube videos. Microsoft operates a Universal Human Relevance System, checking the results of search algorithms with millions of micro-tasks a month. Facebook and Google remediation of ML outcomes was recently intensified by 'cyber-war' scandals (Alba 2017; Dyer-Witheford and Matviyenko 2019). The microwork of data cleaning and monitoring are closely related; both may be performed by the same low-paid contract home-worker on different shifts (Nakashima 2018c).

Such activities illustrate 'the *paradox of automation's last mile:* as AI makes progress, it also results in the rapid creation and destruction of temporary labor markets for new types of humans-in-the-loop tasks' (Gray and Suri 2017). There is a further twist in this paradox: data cleaning and monitoring have themselves become targets for ML automation. Amazon's Mechanical Turk Human micro-tasking is referred to by the corporation as 'artificial artificial intelligence' (Reese and Heath 2016); the irony is that 'turkers' train real artificial intelligences that may one day replace them (Hook 2016). Facebook and Google executives are explicit that their goal is to eventually automate content moderation (Alba 2017; Nieva 2018). Data scientists have declared data cleaning 'a machine learning problem that needs machine learning help' (Ilyas 2018). Click-workers are contributing data sets of their own digitally recorded labour as training material for future AI.

ML thus exemplifies what Tim Jordan (2015: 32–8) terms 'recursion', the process by which information capital feeds into itself. Fear of automation has worked backed to the elite strata of software developers. AI industry workers exhibit a paradoxical relationship to automation in their work; while they are wary of it, they also describe it as necessary and even desirable (Steinhoff 2019b). This then spurs further disagreement as to whether such technologies will enhance or extinguish the work of software engineers (Shanin 2018; Vorhies 2016a, 2017). In 2016, 550 developers were asked to identify the 'most worrisome thing in their careers' from a list of possible concerns: 'I and my development efforts are replaced by artificial intelligence' was the most popular answer (Thibodeau 2016). Some developers dismiss such alarms, believing AI industry growth will only increase demand for human coding and design. Others foresee software developers increasingly assisted by AI 'bots' that automate code checking, distribution and repair, and will increasingly 'spin up' applications in response to natural language commands without developers 'even touching the keyboard', escalating to a point where the 'entire software development lifecycle will be only a high-level description' (Despoudis 2017). Such discussions have been further stirred by the recent BAYOU ML applications, developed at Rice University for the US Department of Defense; this is a neural network that 'trained itself' on millions of lines of human-written code on GitHub, and can reportedly recommend chunks of code to developers on the basis of very brief keyword instructions, thereby taking initial steps towards ML-designed ML systems (Rayome 2018).

The ML production process has a double significance for the analysis of class composition. First, it is an important part of an expanding occupational sector of software development crucial to digital capital. Discussion of this sector conventionally focuses on the well-paid, high-status labour of software engineers and programmers, of whom ML designers are the latest incarnation. This obscures the work conducted 'behind the AI curtain' by data cleaners and monitors (Gray and Suri 2017), part of a global clickwork-force that a UN survey found characterized by 'low pay, poor prospects, and psychological toll' (Berg et al. 2018; Geuss 2018a). All parts of ML's segmented workforce confront a horizon where the very product they create may automate their labour, so that data scientists and data cleaners may both be working themselves out of a job. Second, while ML accounts for only a fraction of global labour, the patterns it discloses of digitally mediated labour, subject to powerful polarizations, and rendered insecure by automation, also appear where the new AI is applied across the wider social factory.

LEARNING THE WAY OF THE HUMAN

The social function of AI as fixed capital in production, like any other labour-automating technology, is the reduction of necessary labour and the concomitant increased extraction of surplus-value. In circulation it accelerates the realization of value, i.e. reduces circulation time, by enabling a tighter integration (coordination and scheduling) between factories, distribution centres and retail stores, and more precise targeting of both seller and advertisers to potential purchasers. It also reduces or eliminates the labour costs of all these activities. In the finance sector, automation speeds the speculative metamorphosis of money into more money. And as AI impinges on consumption and social reproduction, it both eliminates waged workers in industries such as health care, education and entertainment, and establishes new patterns of machine-mediated interpersonal relations (such as social media), or human-machine relations (such as elder-care robots).

How AI is introduced around the planet varies according to capital's spatial organization. In North American and European economies, increasingly dominated by services, retailing and financialization, AI initiatives are strongly oriented towards circulation, although they also attempt to repatriate industrial production. In China, Southeast Asia and other manufacturing zones connected to export markets, AI is strongly oriented to automating industry, though the breakneck speed of economic development also incentivizes circulation-accelerating AIs deployed by companies such as China's Baidu, Tencent and Alibaba. In the sacrifice zones of the planet, the AIs most immediately in play are those of domestic and foreign security forces deploying drones and surveillance systems to control immiserated populations. Here we focus primarily on how AI automation is changing the technical composition of labour in North America and Europe, but taking into account developments elsewhere. Capital's AI deployment is incipient, at a stage of experimental introduction, accompanied by large promises. As Kim Moody points out, in the aftermath of the 2008 crash, business investment in new technologies has been subdued by low profit rates and repressed wage growth; while these conditions persist, the advance of AI automation is likely to be 'bumpy and slow' (2018a: 26). Over the same period, however, corporations have been testing, and using, AI applications in many spheres; as capital's previous waves of technological innovation suggest, this process is likely to escalate if and when capitalists' confidence or workers' wage demands increase. Some prospects may prove far more difficult to implement than others, or even be completely blocked by unanticipated technological or regulatory problems – or, perhaps, working-class resistance. Yet even if only a portion are successful, AI will significantly change the balance of power between proletariat and capital.

In the realm of production, AI in the global North West is hailed as a means of re-industrialization. Manufacturing capacity offshored to foreign locations, or threatening departure, is to be recovered by the injection of robots, sensors and process control computing, all enhanced by ML. While '[m]any industrial AI applications are still somewhat new and bespoke', powerful companies such as GE, Siemens, Microsoft and Intel are 'all making significant investments' in applying ML in manufacturing to bring down labour costs, reduce product defects, shorten unplanned downtimes, improve transition times, increase production speed and conduct predictive maintenance (Faggella 2018b). Siemens has reportedly been using artificial neural networks 'to monitor its steel plants and improve efficiencies for decades' (Walker 2018). In Europe, this project marches under the name of 'Industry 4.0', while in North America it often goes under the broader heading of 'the Internet of Things'.

Such industrial automation is most flamboyantly and problematically exhibited in Elon Musk's Freemont, California factory producing Tesla vehicles (Bhattacarii 2017) – a facility planned not only to produce self-driving cars and trucks but also to itself be so highly automated as to resemble, in Musk's words, an 'alien dreadnought' (DeBord 2017). This craft's operations have, however, been plagued by scandals over injuries to the workers it hopes to dispose of completely (Wong 2017a, 2017b); production bottlenecks Musk concedes were caused by 'excessive automation'; digital sabotage by disgruntled employees; and the crashes and burnings of auto-piloted vehicles that proved better at self-combustion than self-steering. Tesla indicates the difficulties of extreme automation (Wilson and Daugherty 2018). Nonetheless, rising trade protectionism continues to fuel plans for plants tethered to the homeland, even if they employ more patriotic robots than domestic workers. It is, however, an irony dictated by capital's competitive logic that China's corporations, facing labour shortages and wage pressures of their own, are moving in the same direction, and probably faster. Terry Gui's vow to robotize Taiwanese-owned but China-based Foxconn has by no means been completely fulfilled, but nor has it proven totally empty; by 2018 so-called 'Foxbots' were a feature of production lines at Foxconn, as the company expanded its revenues while diminishing its workforce (Chan 2017; Wang 2018). Since 2013, China has been the world's largest market for industrial robots. The majority are purchased abroad, but with growing domestic production, its factories are becoming as or more automated than those in the West (Bland 2016; Ying 2018).

It was, however, in the sphere of circulation that capital's AI initiatives developed most quickly in North America and Europe, accelerating the 'logistics revolution' that has linked transportation, warehousing, communications and retail sales in increasingly complex and seamless systems (Cowen 2014; Rossiter 2016). As we have seen, autonomous vehicles are amongst the most potentially lucrative forms of AI-capital. This prospect is usually discussed in terms of the self-driving personal car (both a consumer commodity and a means by which the commodity labour-power gets to work). The immediate prize, however, is automated commercial trucking. Robo-trucks are already used on-site in industries such as mining. Companies like Volvo, Daimler and Tesla are working on their graduation to long-distance hauling. In 2016, Uber carefully staged a demonstration (under the most favourable conditions) of beer delivery by a self-driving truck (Peterson 2016). Since then such vehicles (accompanied by a watch driver) have entered use in the delivery of refrigerators from a warehouse in El Paso, Texas, along the I-10 freeway, to a distribution centre in Palm Springs, California.

Self-driving trucks, operating 24/7, without rest breaks, in convoys with over-watch, or possibly remote, drivers, on long straight highways free of pedestrians, cyclists and traffic lights or perhaps in special lanes, are a lucrative AI prospect. It would automate an important sector of working-class employment. In the US, trucking is a $676 billion industry, employing about 3.5 million drivers. In a slim-margin business, labour costs 'account for up to 45% of total freight costs' (Campbell 2018). A 2017 report from the International Transport Forum forecast that up to 4.4 million of the 6.4 million truckers in the US and Europe could be eliminated by autonomous technology (ITF 2017). The inclusion of self-driving trucks in national vehicle regulations has been opposed by the Teamsters Union, representing almost 600,000 truckers, which describes their introduction as 'calamitous' (Marshall 2017). Such opposition is only one of a number of barriers facing AI-automated trucking: others include not only continuing problems in refining auto-driving technologies (as witnessed by a series of accidents in 2018), but also the need for adaptions of highways systems and the articulation of legal responsibilities (Moody 2018b). So, while robo-trucks are potentially one of the most profitable of AI-capital ventures, they are also one of the most challenging. There are, however, other more immediately available opportunities.

Less dramatic than cyber-rigs rolling down highways are the invisible software agents transforming capital's circulatory processes. Around the globe, day and night, call centres send outgoing messages soliciting sales for everything from phone plans to museum memberships, and answer incoming calls maintaining, repairing and reinforcing customer relations (Brophy 2017; Woodcock 2017). In Europe and North America, they were promoted as replacing jobs lost to deindustrialization, though such work was then substantially outsourced to India and the Philippines. But as wages for repetitive phone-toil overseas rise, call centres are being AI automated. Customers (and workers) have long been familiar with the recorded message that 'your call may be recorded for training and quality purposes'; that 'training' has come to include training ML systems (Marr 2016). AIs capable of applying natural language processing draw on digital banks of frequently asked questions, as well as knowledge of customers' equipment, competence and past calls to automate answers to common questions, shuttling the more esoteric ones to the remaining human operators, sometimes while simultaneously analysing the callers 'tone, vocabulary, sentiment, and even silences to gauge emotion and satisfaction' (Agrawal, Gans and Goldfarb 2018: 90–1).

The market consultancy company Gartner declared in 2017 that 'customer service spaces' were 'the number one case for applying AI', accounting for 'about 70% of all use cases in AI' (Baraniuk 2018). In 2018, the British retail giant Marks & Spencer replaced 100 switchboard staff with chatbots using Twilio's speech recognition software and Google's Dialogflow AI tool to interpret and route customers' verbal requests. The large US insurer Allstate adopted the digital assistant Amelia – billed as 'the most human AI' – to supplement information staff, reducing call durations from an average of 4.6 to 4.2 minutes – an apparently small saving that across millions of calls adds up to huge cuts in labour costs. Call companies like to term such changes 'augmentation' rather than 'automation' of human staff. For the moment, its main effect may be to intensify the work of human operators, but many observers believe automation will prevail in the long term (Wood 2018). Similar capacities are being extended to other corporate communications. In 2018, Google demonstrated how its Duplex system combines calendar search and natural speech patterns to schedule appointments by phone, with voice simulation difficult to distinguish from a human; it was promoted as 'useful for the 60% of small businesses in the US that don't already have an online booking system' (Solon 2018).

Automated transport will move goods to and from warehouses that are themselves already automated by AI. Amazon's 'fulfilment centres' are the paradigm case. Amazon is a company that has made ML its 'flywheel' (S. Levy 2018) and has deployed it in the corporation's algorithmic recommendation system, online purchases, smart home devices and cloud computing empire. It is also in the cavernous fulfilment centres where orders are received, picked from shelves, packed, and loaded to trucks, by notoriously poorly paid, speeded-up, stressed-out and electronically monitored workers. Demonstrating Marx's thesis that capital's machinery intensifies rather than diminishes exploitation, the heightened pressure on human 'ambots' has accompanied the deployment of real robots (Greenmeier 2008; Abdelrahman 2017). Amazon acquired robotics company Kiva Systems (renamed Amazon Robotics) in 2012 for $775 million, inspired by the discovery that 'fulfillment center workers were spending more than half their time walking around the warehouse to find items and put them in their tote' (Agrawal, Gans and Goldfarb 2018: 145). Large-scale deployment of its round, low automata followed a Christmas crisis during which the corporation failed to meet promised delivery dates. By 2017 Amazon had more than 100,000 Kiva bots in select fulfilment centres worldwide (Wingfield 2017). Where these bots are used, they are directed by a constantly updated computerized inventory system and guided by motion sensors to bring goods to pickers stationary on a work platform. The robots have reduced the time to complete an order by a fifth. They save Amazon space – their squat profile allows fitting 50 per cent more inventory in a given area – and power bills; Kivas work in the dark and do not need air conditioning to prevent the heatstroke that regularly brought ambulances racing to fulfilment centres (Tam 2014).

Amazon's fulfilment centre automation still has major gaps. One is the 'picking' of items from the robot-propelled mobile shelves. This may seem a routine task easily mechanized, but an Amazon warehouse, unlike an auto assembly line, deals with an 'almost infinite variety of shapes, sizes, weights and firmness of items' (Agrawal, Gans and Goldfarb 2018: 144). This has made picking an intractable automation problem. For several years an Amazon Picking Challenge invited major robotics companies to find a solution, without success. One recent line of research uses a 'a mix of automated software and a human controller', with the robot automatically navigating to a shelf where the human – wearing a virtual reality helmet, possibly in a remote location – guides its arm to grip and move the item; the long-term aim is to use a machine-learning artificial intelligence 'trained on many observations of human grasping via tele-operation to teach the robot to do that part itself' (Agrawal, Gans and Goldfarb 2018: 145).

Other companies may overtake Amazon. The UK online-supermarket Ocado is a leader in designing highly automated systems it sells to other grocery chains. Its warehouse system looks like a 'a huge chessboard, populated entirely by robots' that engage in tasks of 'lifting', 'moving' and 'sorting' in a 'hive-grid-machine'; each bot has a central cavity and claws to grab and pull crates inside 'like an alien abduction in a supermarket aisle' (Vincent 2018). The bots are interchangeable, switching from one task to another; production can be scaled up or down by the addition or subtraction of bots, and their algorithmic programming allows novel solutions to warehouse organization. China's warehousing firm JD.com, working in collaboration with Google, claims to have achieved even higher levels of automation (Paquette 2018). There are still plenty of workers in Ocado and Amazon facilities, but as bot systems deepen and spread, the pace of work increases and its security declines.

Circulation culminates in sales. Retail jobs too are registering the tremors of AI's introduction. Again, ML builds on previous digital changes; online sales have transformed retail business, self-checkout lines are common, and both made deep cuts to in-store employment. AI extends this logic to the creation of staff-less automated retail outlets run almost entirely on downloaded apps and scanned products. These technologies were pioneered in Sweden, initially in remote-area grocery stores (Prindle 2016). They were then deployed by Swedish companies in China (Sun 2017), where mobile phone payment for daily shopping has been popularized through services such as WeChat, and no-staff stores, some operating on a relatively low-tech self-checkout basis, are common. The North American version debuted in Amazon's Go stores which, as we have discussed, combine mobile phone apps, automatic scanning and 'artificial intelligence and computer vision to match the face of the buyer with the items in one's bag, to eliminate checkout altogether' (Horwitz 2018). Such systems have the potential to include chatbots providing shoppers with 'customized suggestions', and IoT delivery orders from automatically self-stocking refrigerators and other smart home devices (Sun 2017; Pirrone 2018). Walmart, already famous for its massive system of computerized inventory control, is researching similar systems in partnership with Microsoft, a development that, given Walmart's status as the largest single corporate employer in the US, has major labour market implications.

Shopping requires money, and in an era characterized by the expansion of both debt and speculation, financial institutions have been amongst the earliest adopters of AI. ML methods are now widely used 'to assess credit quality, for fraud detection to price and market insurance contracts, and to automate client interaction' (FSB 2017). But it is the highest levels of the finance sector that provide the most striking examples of AI automation. At the time of the Wall Street meltdown, stock traders were already being decimated by the algorithmic agents of high-frequency trading. Since then, stock exchanges in Tokyo, Singapore, London and Hong Kong have eliminated their trading floors. By 2018 only some 10 per cent of stock market trades were actively conducted by humans; 40 per cent are 'passive' trades scheduled by mutual funds; the remaining 50 per cent are executed by algorithms that are increasingly informed by ML (Pearlstein 2018; Coles 2018). In 2000, Goldman Sachs equity trading employed some 600 people: by 2016, all but two had been replaced by 'complex trading algorithms, some with machine-learning capabilities' (Byrnes 2017). J.P. Morgan revealed plans for an AI focus (Terekhova 2017), and poached an ML expert from Microsoft to oversee this reorganization (Kolanovic and Krishnamachari 2017). The one-time 'masters of the universe' were being eaten by the tools they had helped create (Byrnes 2017).

Descending from the heights of financial manipulation to the depths of everyday scraping-by, we arrive at the role of AI in social reproduction. Algorithmic profiling is ubiquitous in corporate and state decisions on loans, insurance, medical claims and job applications. It is applied with special ferocity on those who fall on the margins or outside of waged labour, i.e. the reserve army of the unemployed and the surplus populations likely to be enlarged by AI automation. As Cathy O'Neill demonstrates (2016), such machinic decisions, by intention or oversight, persistently discriminate against all indications of penury and precarity, whether by postal code, ethnicity, gender or credit history. Virginia Eubanks (2017) documents how the use of automated decision-making tools to administer welfare benefits, social housing and family care interventions constructs an AI managed 'digital poorhouse' that denies or deters applications, subjects recipients to police scrutiny, and stigmatizes them on the basis of behavioural patterns and past associations. A corollary of such control is the well-documented (Larson et al. 2016; Lum and Isaac 2016; Ensign et al. 2017) deployment of AIs for predictive policing and sentencing, promoted as reducing human bias, but in fact algorithmically continuing and amplifying previous patterns of judicial and police racism in a 'runaway feedback loop' (Reynolds 2018). This is the sharp end of the more general process by which ML, but specifically deep learning (Impett 2018), becomes in the hands of capital the means for an ever more rigorous, comprehensive and recursively self-confirming sifting, segregating and disciplining of various grades of permanent, precarious and surplus labour-power – becomes, that is, an instrument of proletarian decomposition.

APOCALYPSE NOW OR BUSINESS AS USUAL?

Marx, as we have seen, understood technological unemployment as a weapon in the conflict between capital and labour. The issue of AI automation's effects on employment is central to the analysis of class composition. Controversy about this topic has raged amongst bourgeois economists and technologists for the last decade. But unlike Marx, none of the participants in this mainstream debate want capitalism destroyed. On the contrary, all are all anxious to preserve it. The advocates of what we will call the 'AI Apocalypse Now' position believe an imminent jobs crisis calls for emergency measures to save the system, while 'Business-as-Usual' proponents point to historical precedent to claim that labour markets will adjust to AI with only relatively minor employment shocks.

The AI Apocalypse version is often presented by futurists with a computer science background, such as Martin Ford (2009, 2016). He acknowledges that capitalism has sustained employment through successive waves of technological change, from steam engines to electrical power, and early computerization. Nevertheless, Ford asserts that AI and other fourth industrial revolution technologies are different because of their exponential, Moore's-Law-driven speed of improvements and cross-sectoral applicability. Previous technological transformations moved slowly from one industry to another. AI automation, Ford says, will have a near simultaneous adoption; workers displaced from, say, 'Industry 4.0' manufacturing will be unable to find work in call centres because these too will be AI automated. MIT computer scientists Erik

Brynjolfsson and Andrew McAfee (2014) make a similar argument, emphasizing how digital automation now extends beyond routine, manual 'blue collar' work to 'white collar' occupations such as journalism, advertising, lawyering, medicine and other middle-class professions. For capital to survive, radically new social policies would be required, including the introduction of a universal basic income or citizen's wage (an idea we discuss in our conclusion to this book). Such prophecies seemed corroborated by a 2013 study from Oxford economists Carl Frey and Michael Osborne, suggesting that 47 per cent of the 702 occupations into which US jobs are conventionally sorted are 'likely to be substituted by computer capital' within two decades. Coming amidst AI exploits like Watson's *Jeopardy* victory and the defeat of grandmaster Lee Sedol by Google's AlphaGo, this study helped make the hailing of future robot overlords a virulent cultural meme.

Against this, the Business-as-Usual position is championed by professional economists (Autor 2015). It usually suggests that the achievements of narrow AI are mainly in formalized and/or highly predictable situations. These limitations, alongside the regulatory problems confronting technologies such as self-driving cars, will, Business-as-Usual theorists argue, slow AI adoption, making a sudden-onset, across-the-board crisis unlikely. More importantly, while automation destroys some jobs, it creates others. Some emerge in the very industries that build automating technologies, such as the data scientists and microworkers who make AIs. But other new employment opportunities emerge even in the industries AI automates. So it is suggested that autonomous vehicles may make truckers and taxi-drivers obsolete, but will create jobs for engineers designing hardware and software, as well as for workers 'taking customer calls, cleaning and repairing cars, and updating ... high-definition maps' (T. Lee 2018). To stay competitive, firms will have to pass some benefit on to consumers in the form of lower prices; this will increase real income levels, enhance consumption and raise employment. The 'displacement effect' of automation – workers being thrown out of jobs by AIs – will be offset by the 'income effect' of heightened demand for labour (Hawkesworth, Berriman and Goel 2018). Business-as-Usual advocates admit the possibility of a 'difficult transition' and reiterate that 'there will be winners and losers', but the overall message is that AI-driven job loss will be limited, and remediable by skills upgrading: capitalism should keep calm and carry on.

The argument, raging for several years, has become well-choreographed. In 2018, the BBC reported on a study conducted for the City of London that suggested a third of jobs in the UK capital would be performed by machines within the next 20 years; wholesale and retail sales, transportation, storage, and food industries, employing a million people, would be hard hit (Kulka and Brown 2018). The Bank of England's chief economist, Andy Haldane, interviewed on the topic, declared that the 'Fourth Industrial Revolution' would be of a 'much greater scale' than changes wrought by steam power, electricity and early computerization. Though these had 'a wrenching and lengthy impact ... leaving many out of work for long periods struggling to make a living', such effects would be yet more severe 'when we have machines both thinking and doing – replacing both the cognitive and the technical skills of humans' (Morrison 2018). Immediately, Simon Jenkins (2017), a reporter with *The Guardian*, responded with an article: 'Worrying About Robots Stealing Our Jobs? How Silly'. Noting a chronic labour shortage in British social services such as health and education, Jenkins declared fear of AI mass unemployment absurd. New technology would mechanize some activities, but, 'as throughout history', labour markets would shift and 'people – or their offspring – retrain'. 'Short term disruption' was 'the reality of economic history', yet 'we survived and prospered as new needs, and jobs, emerged'. Another industrial revolution would release time for work in 'service industries [that] essentially involve human relationships' and 'cannot be done by robots or machines'. Economists would, Jenkins concluded, do better to 'welcome AI as releasing workers into the experience economy'.

Since the two contending positions were first outlined there have been a spate of predictive studies. Some are produced, commissioned or published by elite advisory organs of capital, such as the Organization for Economic Cooperation and Development (OECD 2016; Arntz, Gregory and Zierahn 2016; Nedelkoska and Quintini 2018) or the World Economic Forum (WEF 2016, 2017); others by private business consultancies with an interest in AI-related opportunities, like the McKinsey Institute (Manyika et al. 2017), PricewaterhouseCoopers (PwC 2017; Rao and Verweij 2018; Hawksworth, Berriman and Goel 2018) and Gartner (2017). Others come from legal associations dealing in labour law (Wisskirchen et al. 2017); the International Labour Organization (Chang and Huynh 2016) and anti-poverty organizations (Boston Consulting Group–Sutton Trust 2017). There have also been many independent academic studies.[1] This outpouring has in some respects refined analysis and in others rendered it more confusing.

In general, studies that focus on AI replacing entire jobs – such as that of Frey and Osborne (2013) – have been succeeded by more fine-grained analyses of tasks within any one job that are susceptible to automation. This allows for possibilities of AI complementing or intensifying human labour rather than entirely substituting for it. In the preferred business jargon: 'augmenting' rather than 'automating'. Subsequent OECD studies have dialled back the percentage of employment at imminent risk of automation from the apocalyptic to the merely alarming: 14 per cent for all its member countries, equivalent to over 66 million jobs, with another 33 per cent 'deeply restructured' by automation. In the US a mere 9 per cent or some 13 million jobs are tagged as AI-vulnerable, although the authors remark 'this would amount to several times the disruption in local economies caused by the 1950s decline of the car industry in Detroit' (Nedelkoska and Quintini 2018). The picture is further revised when various estimates of employment opportunities created by AI are reckoned in. A few have gone so far as to promise overall employment increases consequent on technologies of the fourth industrial revolution. Yet they often still admit the possibility of a net job loss (Arntz, Gregory and Zierahn 2016; Economist 2016a; Nedelkoska and Quintini 2018; Acemoglu and Resteropo 2018).

There have also been reversals over the types of jobs most liable to replacement. One of the main claims of AI Apocalyptics (and, one might cynically suggest, a reason for the prominence given to the issue) is that not only manual but also mental 'middle-class' jobs are threatened. More recent studies nevertheless stress that it is routinized work most immediately at risk. It is generally agreed that transport drivers, administrative assistants, cashiers, counter and rental clerks, telemarketers and accountants are the most endangered, and of those middle-class jobs, the most likely to be automated are 'paraprofessional' jobs, such as 'paralegals, payroll managers, and bookkeepers' and 'semi-administrative jobs' that provided a way into professional industries for those without advanced educational qualifications (Vincent 2017).

The geographical distribution of AI-induced automation is similarly contested. Many studies assume that developed economies are likely to be hit hardest, as 'higher average wages incentivizes automation' (Manyika et al. 2017). Others suggest AI may have the most dramatic effects in less advanced areas of the planet factory, precisely because there are more routine industrial jobs there (Kinder 2018). A study by the International Labour Organization (Chang and Huynh 2016) estimates that more than 137 million workers in Cambodia, Indonesia, the Philippines, Thailand and Vietnam are at high risk of replacement by machines such as 'sewbots' that combine automated sewing with computer vision (Larson 2018). Such 'premature deindustrialization' (Rodrik 2015) jeopardizes models of capitalist development, threatening to 'slash millions of jobs and create an upswing in trafficking and slavery across south-east Asia' in a 'race to the bottom' between the costs of machines and of humans (Kelly 2018). The geographical distribution of job losses in the centres of advanced capitalism will also be uneven. A study by Frank et al. (2018), focusing on the US, argues that job losses to AI automation will affect small cities worst because large cities have a disproportionate number of occupations that are of a cognitive or analytical nature at less risk of automation, while small city jobs are disproportionately routine, such as clerical work in retail and food service.

The great speculative AI and jobs debate has been coloured by changing economic conditions. AI Apocalyptics had a major impact in the high unemployment aftermath of the 2008 crash. When, a decade later, recovery was declared and employment levels restored (at least in the US), Business-as-Usual pundits regained their confidence. The tone of most recent managerial reports and policy advisories on the traumatic effects of the fourth industrial revolution is neither 'denial' nor 'doomsday' but 'daunting', with discussion of major Marshall Plan-sized projects for retraining workers. Great uncertainty thus continues to dog the AI and jobs debate. Behind the euphemisms and promises, however, there is visible the likelihood of a world of proletarian woes, with large sectoral job losses and chronic insecurity. And even in the most optimistic estimates of new AI-related job creation, there remain other important questions about its consequences for class relations, to which we now turn.

PANOPTICS, PRECARITY, POLARIZATION AND PROLETARIAN SCHOOLING

As Valerio de Stefano (2018) has pointed out, discussions about AI-related job loss and creation focus on quantitative rather than qualitative employment issues. While the size of the reserve army of the employed affects class decomposition and recomposition, so too do management techniques, the space and temporalities of labour, divisions within the workforce, and the forms of training and education through which that workforce must reproduce itself. As one commentator remarks, it might be true that 'robots will create more jobs, ... but what if these jobs are less good and less well paid than the jobs that automation kills off?' (Elliott 2017). We consider four points: panoptics, precarity, polarization and proletarian schooling.

The cover of an *Economist* (2018) report on 'Artificial Intelligence in the Workplace' depicts a Brobdingnagian office lamp illuminating Lilliputian workers, under the title 'AI-spy'. Digital technologies have brought a 'new Taylorism' (Salame 2018) to workplaces, in which job deskilling is fused with algorithmic management and accompanied by workplace surveillance capacities far greater than those of any clipboard carrying superintendent. Phoebe Moore and her colleagues (2018) detail how sociometric badges, keyboard counters, email scanning, location tracking, motion sensors, and voice and facial recognition technologies detecting shifts in efficiency and mood minute to minute are now normal in workplaces. ML-driven 'People Analytics' is the new obsession of corporate HR departments that draw on AI tools for employee screening and selection based on social media traces, and delegate interviews to chatbots (Buranyi 2018). Fed with such data, AI becomes a way of assessing workers' 'productivity and fitness to execute particular tasks' (De Stefano 2018: 7). This creation of a digitally 'quantified worker' is advanced under the guise of providing employees with beneficial information about their performance, health and state of mind for the purposes of voluntary self-improvement and 'wellness'; the corporate providers of wearable workplace devices market them as 'humanizing' technologies. But this barely masks their obvious enhancement of managerial power by such technologies of 'limitless worker surveillance' (Ajunwa, Crawford and Schultz 2017).

Primarily concerned with the stress, indignity and intrusiveness such surveillance inflicts on employees, Moore et al. (2018) are sceptical about 'deep automation' and emphasize the economic, technological and social barriers to its abrupt arrival. But such monitoring is a moment within AI development. By the same logic that AlphaGo's defeat of grandmasters at the ancient game of Go depended on digital digestion of thousands of hours of previous human play, so recording worker performance, even if initiated for disciplinary reasons, opens the way to new transformations of labour into machine form. Autonomous vehicles depend on digitally recording hundreds of thousands of hours of human driving. The recordings of call centre staff become the basis for voice analytic software that first supplements their work and then supplants it. The robotization of Amazon fulfilment centres emerged from electronic tagging of human 'ambots': in 2018 the company had two patents approved for a wristband tracker emitting ultrasonic sound pulses and radio transmissions to monitor a picker's hands in relation to inventory, providing 'haptic feedback' to 'nudge' the worker towards the correct object (Yeginsu 2018). Reports of these patents sparked a public outcry about worker privacy and speed-up, but it is also clear that the wristband, or even more extensive monitoring devices, could pave the way for the robotic replacement of pickers. The organic link between workplace monitoring and worker replacement, and hence between AI development and surveillance, led *The Economist* (2018) to conclude that while AI workplaces may not be immediately depopulated, they will become 'creepier'.

Precarious employment is one of the most discussed features of work in twenty-first-century capitalism. Definitions of 'precarity' vary, but often include part-time, temporary and self-employment, widely thought to be on the rise since the 2008 crash (Standing 2011). Particular attention has been given to the growth of a 'gig economy', organized by online platforms such as Uber, Lyft, TaskRabbit or UpWork, Clickwork and Mechanical Turk by workers designated as independent contractors, hence working without pensions, benefits and other regulatory protections, and often for low wages (Hill 2015; Kessler 2018). Some analysts of labour politics argue that the scale of 'gigging' is exaggerated (Henwood 2018b; Moody 2018b). In this view the real problem for workers today is not so much precarity, but rather stagnating low wages, rising costs and poor conditions in the permanent jobs that still account for the largest part of employment. These two perspectives are not entirely contradictory, as workers may supplement low-wage permanent jobs with precarious employment.

There is an affinity between AI and precarious labour. As we have seen, AI depends on online microwork. In addition, AI is important for the algorithmic management used by platform capitalism to organize contingent workers, scaling their numbers, earnings and schedules to demand, and thus 'contributing to a casualization of work patterns and job and income instability' (De Stefano 2018). Uber is a company self-described as having 'machine learning in its DNA' (Reese 2016a).

When it was rapidly expanding in 2014 and 2015, Uber was adding up to 50,000 drivers every month, a number impossible for human managers to schedule and supervise, 'so some system of virtual management had to be put in place' (Caddell 2017). The company filled the gap with a variety of automated management tools providing a 'rating system, performance targets and policies, algorithmic surge pricing [and] insistent messaging and behavioral nudges', that 'steer drivers to work at particular places and at particular times' – a system that, as Alex Rosenblat (2016) remarks 'complicate[s] claims that drivers are independent workers whose employment opportunities are made possible through a neutral, intermediary software platform'. From 2016 Uber consolidated these algorithmic management tools within its own internally produced 'Michelangelo' ML system, which was 'designed to cover the end-to-end ML workflow: manage data, train, evaluate, and deploy models, make predictions, and monitor predictions' (Hermann and Del Baso 2017). The logical culmination of Michelangelo's trajectory seems to await the fulfilment of Uber's ambitions to supply the software for self-driving cars, in which case the data culled from hundreds of thousands of precarious self-employed drivers would be absorbed by AI-directed autonomous vehicles.

A third current tendency AI is likely to amplify is the polarization between 'high-end' and 'low-end' jobs. This trajectory, apparent in the US since the 1950s, has intensified as digital automation and outsourcing hollowed out 'middle-level routine' jobs in favour of either high-paying cognitive labours or low-paying manual work, albeit with downward pressures on both (Autor and Dor 2013; Beaudry, Green and Sand 2013; OECD 2016; Elliott 2017). The US Department of Labour's (2015) occupational projections to 2024 lists 'computer and mathematical occupations' as one of the fastest growing, and best paid, areas of employment. AI-capital will (at least until its self-automating capacities kick in) contribute to this growth, as it requires not just direct programming of ML, but also coding of the social media, apps and sensors that provide the Big Data on which it depends. The US national average salary for IT jobs is about $81,000 (more than double the national average for all jobs), and the field is 'set to expand by 12 percent from 2014 to 2024, faster than most other occupations' (US Department of Labour 2015). But coding labour is dwarfed in absolute numbers and rate of growth by jobs in areas such as personal care work, nursing and medical assistance, retail sales and food service, many with wages in the $20,000 a year range

(US Department of Labour 2015; Henwood 2015). This suggests a future labour force divided between programmers coding other mid-level jobs out of existence, leaving tasks such as the care of an aging population to workers whose jobs remain at once too cheap and too humanly complex to automate – though this does not inhibit rampant fantasies about the imminent possibility of sex-bots and robot teachers. Such dreams and nightmares may be finding realization in the elder-care facilities of Japan, where the demographic problems of an aging population are particularly acute, and automata find relatively easier cultural acceptance. The development of 'social robots' as potential aides to caregivers has been a focus of government-supported industrial research (Kyung-Hoon 2018), and the Japanese public is exposed to the constant promotion of 'exoskeletons, internet-of-things gizmos, humanoid or animal-shaped robots ... destined for a career of endlessly cheerful toil in the country's growing network of nursing homes' (Lewis 2017). Elsewhere, such trajectories will probably be slow-moving, at least while wages for personal care workers remain low.

This polarization has gendered and racialized dimensions. Early waves of digital automation hit men particularly hard because of their effects on manufacturing employment: the 'information revolution' is often cited as a contributory cause to the gradual advance of women in many fields of work. Many of the jobs likely to be AI automated are, however, typically held by women in administrative, secretarial, sales and customer service work, while software engineering is masculinized (Howcroft and Rubery 2018). In North America, the number of women studying computer science is static or falling, while in the UK only one in five of computing students is female. It seems likely that 'men will disproportionately benefit from the greater demand for STEM skills' (Boston Consulting Group–Sutton Trust 2017: 21–2). Silicon Valley companies have demonstrated an incapacity to move beyond a 'brogramming' culture, and also display ethnic segregation: a report from the US Equal Employment Opportunity Commission found that just 8 per cent of tech sector jobs were held by Hispanics, 7.4 per cent by African Americans, and 36 per cent by women (2016). Such patterns may be intensified by algorithmic hiring tools. In 2018 it was reported that an Amazon AI recruiting technology trained predominantly on men's résumés had unsurprisingly developed an intrinsic bias against women, having 'effectively taught itself that male candidates were preferable' (BBC 2018b). That the system was discarded reflects a growing awareness of and resistance to high-tech sexism, but it is not an isolated example; analysis of automated hiring processes by both LinkedIn (Reese 2016b) and Google (Gibbs 2015) suggest similar sexist biases. At the other end of the spectrum, many of the fastest growing low-wage service jobs are ones in which women and minorities are strongly represented. AI-capital may therefore mark a new round of gendered and racialized work segregation.

The route to survival in the face of AI automation is universally held to be education. Promises of 'reskilling' for new AI-related work can be glibly made at the macro-level of the entire workforce, but the micro-fate of individual workers will not be so easy. A capitalist nostrum is that the technologically unemployed should upgrade their programming skills and computer literacies. In the recessionary, post-2008 crash context of abundant low-cost labour, corporations widely abandoned training: it was cheaper just to hire someone new with the relevant skills. In such a labour market, workers and the unemployed have to conduct costly upgrading efforts at the very time when neoliberal policies have made access to education sharply stratified. As Dan Shewan (2017) writes of the US:

> Private schools such as Carnegie Mellon University ... may be able to offer state-of-the-art robotics laboratories to students, but the same cannot be said for community colleges and vocational schools that offer the kind of training programs that workers displaced by robots would be forced to rely upon. In light of staggering student debt and an increasingly precarious job market, many young people are reconsidering their options. To most workers in their 40s and 50s, the idea of taking on tens of thousands of dollars of debt to attend a traditional four-year degree program at a private university is unthinkable.

Notional remedies for this problem include 'massive open online courses, career-oriented nanodegrees ... that provide industry-recognized credentials, and coding boot camps that teach career-ready IT skills in a few months' (Kinder 2018). At present, however, 'these programs are used predominantly by those who are already highly educated or digitally savvy and looking to further enhance their employability by mastering cutting-edge technology. Few workers at the lower end of the labor market are taking advantage of such programs' (Kinder 2018). These problems will be intensified because in an AI-suffused context 'the speed of technological change will require these skills to be acquired rapidly, but they will also become obsolete faster with a shorter "half-life", so that 'the need to continuously re-skill and up-skill oneself will raise the cost and time required for education and individual development' (Boston Consulting Group–Sutton Trust 2017). Regardless of the net loss of jobs associated with AI, the concatenation of problems it brings around issues of precarity, labour market polarization and job restructuring threaten greater difficulties in workplace organizing, intensified divisions between the new elites of high-technology labour and low-wage or unemployed workers, and for most people compounding stresses and costs around education and training. It will also more widely affect how people think, communicate and interact around political issues, a point we address in the following section.

GRAMMATIZING THE NETWORKS

It is intrinsic to the concept of the social factory that, though composition and decomposition of the working class may start in the workplace, it extends beyond it (Notes from Below 2018a). Within the factory gates, the working class was decomposed by Taylorism and by increasingly advanced automation technology. This process of deskilling continues with AI, but increasingly extends into the spheres of circulation, consumption and social reproduction. To explain the effects of the AI-automated social factory we borrow the concept of 'grammatization' from Bernard Stiegler (2010, 2015, 2018), precisely because he uses it to refer to a deskilling that affects workers' knowledge not just of how to work but also of how to live and think, a process he relates directly to AI.

For Stiegler, grammatization is 'the process through which the flows and continuities which weave our existence are discretized' (2010: 31). This concept refers to the development of technologies that record, separate and codify aspects of human activity. The translation of speech into alphabetic writing, with its grammatical rules, is an early and preeminent example of this process (hence 'grammatization'). But grammatization, in Stiegler's usage, also involves sense organs (vision and hearing), the movements and gestures of the body (at work, at home or while consuming), and entire patterns of social interaction (Stiegler 2010: 10; 2018: 8). The deskilling of workers by the transfer of their knowledge to machines in a way that subordinates them to capital is but an initial moment of a widening process of the grammatization of proletarians (which we gestured to in the previous chapter). This process later spreads to the automation of consumption – people are trained as consumers by the audio-visual technology of the culture and advertising industries – and then to human self-understanding, a knowledge that is automated by, for example, social media and big data analytics.

Grammatization, we argue, should be understood as an extension of the real subsumption of labour to what Marx in the *1844 Manuscripts* termed 'life-activity' (2007: 75). What ML subsumes is not just labour but more general human behaviour, communication, knowledge and skills that relate to how we live – including purchasing and consumption, but also the social relations of friendship, conviviality and love that are, from capital's point of view, engines to be harnessed for its accelerated circulation. As Stiegler (2010: 45) puts it, using three useful French phrases, *savoir-faire* (knowing how to make and do), *savoir-vivre* (knowing how to live) and *savoir-être* (knowing how to be) are stripped from humans across the entire span of the social or planetary factory. In a recent work he explicitly relates this to AI and big data, suggesting that in their current forms these create disaffected, confused and incapacitated subjectivities, in a profound sense 'stupefied' (Stiegler 2015). For Stiegler, machine learning 'calculate[s] correlations ... in order to automatically anticipate individual and collective behaviour, which they also provoke and "auto-realize" by short-cutting and bypassing any deliberation' (Stiegler 2017: 231). In our terms, they create an intensified state of decomposition.

It is through social media that capital is exercising some of its most powerful AI interventions, which at once restructure everyday social relations and have major implications for class composition. Google's search-ranking algorithm and Facebook's social graph of networked personal relationships have for over a decade been transforming the activities of these platforms' users into digital profiles for targeting content and sale to advertisers. Where this crosses the boundary into the realm of AI is debatable; algorithmic processing is subject to the AI effect. Nevertheless, across two significant frontiers – the capacity to improve their own performance, and the escalating degree of predictive confidence that results – both corporations claim to be taking giant strides.

Facebook in 2016 unveiled its 'self-improving, artificial intelligence-powered prediction engine', dubbed 'FBLearner Flow (FBL)' (Biddle 2018). This uses 'machine learning expertise' to draw on 'location, device information, Wi-Fi network details, video usage, affinities, and details of friendships, including how similar a user is to their friends', to address corporate 'core business challenges'. FBL is not merely offering advertisers the ability to target people based on demographics and consumer preferences, it also provides a 'loyalty prediction' service that searches a user base of over 2 billion individuals to detect millions 'at risk' of jumping ship from one brand to a competitor, subjects primed to be 'targeted aggressively with advertising that could pre-empt and change their decision entirely'.

Another example of capital's AI-powered grammatization of personal communication is Google's Federated Learning project. This was announced in 2017 as a 'decentralized' edge AI, whereby 'machine learning models are trained directly on smartphones of users'. Google could now delegate AI training to Android phones via an app, which would read users' files, improve the personalization of their digital profiles, and relay this information to advertising clients, all the while leaving the original data 'intact' on the phone rather than lodging it on Google's servers. Although Google chose to feature the user privacy benefits of this approach, Federated Learning brought the corporation a number of other, more self-interested, advantages. 'Decentralized' AI that functions 'locally' on users' devices with no dependence on a network connection, 'means less power consumption, ... minimal latency and faster machine learning processing', and actually gives access to '*more* user data' than could practically be harvested to a remote server in the cloud (Kulian 2017, emphasis added).

In the wake of the 2008 financial crash, it seemed that the growing reach of corporate-owned social media might actually expand possibilities for social dissent. The wave of occupations, assemblies and riots that burst out in 2011 were widely dubbed 'Facebook (or Twitter, or YouTube) revolutions'. Although this appellation has been widely and properly criticized for its absurd techno-determinism (e.g. Tawil-Souri 2012), it remains true that a feature of these uprisings was the extensive use of social media platforms and mobile phone networks by protestors (albeit alongside good old fashioned on-the-ground organizing and face-to-face communication). This contributed to the rapid mobilization and contagious spread of protests, though probably also to their later problems of political coherence and short staying power (Dyer-Witheford 2015).

In the aftermath of these unrests, however, two AI-related developments posed major problems for networked anti-capitalist movements.

The first was the discovery of the scope of nation states' digital surveillance of populations, and in particular Edward Snowden's disclosures about the US National Security Agency (NSA). These revealed that surveillance programmes set up in the wake of 9/11 piggy-backed on the algorithmic data collection of US social media and search engine corporations, all of whom collaborated with the NSA's PRISM programme. What Snowden also showed was that the NSA processed the information gathered from PRISM and other projects by means of its own advanced AI programmes, collecting data and metadata, storing it on NSA cloud servers, and applying ML to identify suspect activities and associations (Gallagher 2013a, 2013b; Grothoth and Porup 2016). While the ostensible target of such AI-driven surveillance was terrorism, its technologies could unquestionably be applied to other purposes. Protestors already subject to 'cyber-crackdowns' via CCTV, mobile phone and social media surveillance were now potentially subject to a new extreme of scrutiny, if not necessarily from the NSA itself, then by the downward creep of AI-driven surveillance from elite intelligence agencies such as the NSA and GCHQ to police forces, private security firms and corporate labour spies (Dyer-Witheford and Matviyenko 2019).

The second pulse of AI-driven decomposition into internet politics can be summed up in the name 'Cambridge Analytica', if we take this as a shorthand for the convergence of alt-right networks, electoral digital strategy and Russian cyber-war interventions around Donald Trump's election campaign. The contribution of AI to this scandal, which is still ramifying, has several levels. First, while the details of the voter profiling and targeting programs used by billionaire Robert Mercer's company to assist Trump remain opaque, his background as an AI pioneer make it a virtual certainty these involved ML. Second, the Trump campaign relied on Facebook, an ML pioneer, both for the data obtained by Cambridge Analytica, and for the overt assistance Zuckerberg's company provided Trump's campaign in tailoring its digital advertising. Third, Trump's electoral strategists, his alt-right supporters and friendly Russian operatives all relied for the circulation of their messages on social media algorithms promoting provocative, attention-grabbing, and hence advertising-lucrative content. By 2016, these algorithms were engineered with the assistance of ML. It is not for nothing that Robert Mercer's daughter, Rebekah, devised a parlour game based on Trump's campaign called 'The Machine Learning President' (Mayer 2018).

Trump's election was an exemplary instance of decompositionary class politics: it aimed precisely at dividing people along lines of race and gender – with special attention to rust-belt populations damaged by globalization – in order to get them to vote for a capitalist plutocrat. Trump is, and will be, far from the only political figure to deploy ML technologies; Obama was, after all, 'the big data President' (Scola 2013). His campaign, his supporters and similar ventures such as the UK Brexit campaign have demonstrated how deeply AI-powered interventions can corrupt network communication. They have also shown the affinity between right-wing populism, or, more properly, neo-fascism, and digital exacerbations of the divisions within working classes already fragmented by decades of neoliberal class warfare, and on the verge of fresh AI-driven fractures. At the very moment when recompositionary initiatives around the social factory seem most important for anti-capitalist politics, these have been rendered far more difficult by the attention-shattering impact of algorithmic advertising, the chilling effect of ML-informed mass surveillance, and inflammatory fake news, toxic chatbots, cyber-warfare and other forms of 'weaponized AI propaganda' (Anderson and Horvath 2017). AI thus contributes to the transformation of the internet from a potential arena for the 'circulation of struggles' (Dyer-Witheford 1999) to one dominated by the circulation of commodities, the surveillance of resistances and the destruction of class solidarities.

A HEPTAGON OF STRUGGLES

Nonetheless, popular apprehension about the scale and direction of capitalism's AI ambitions is growing. Recent years have seen resistances and protests relating to AI springing up around the social factory. What we emphasize is that these are eruptions of popular sentiment, not for more AI ('full automation now'), but for its *refusal*. None are outright anti-AI struggles, but each rejects or contests specific aspects of AI. In their combination and overlap, and despite frequent disconnections and contradictions, they challenge the current trajectory of AI-capital. We list seven of these antagonisms – a heptagon of struggles.

1) Strikes and other workplace actions against AI-driven wage-depression, speed-up, monitoring, precarity and algorithmic management. The preeminent example is the wave of strikes and walkouts across Amazon fulfilment centres in Germany, Italy, Spain, Poland and the UK, demanding wage increases, job security and improvements in worker safety (Boewe and Schulten 2017; Cillo and Pradella 2017; Amazon Workers and Supporters 2018). The October 2018 pledge by Jeff Bezos to raise Amazon workers' minimum wage (though undermined by claw-backs to a variety of other payments) was a partial victory for this movement, a concession intended to head off any transatlantic migration of struggles from Europe to the US (Reese and Struana 2018). Other strikes have disrupted the algorithmic management of precarious labour. The food courier Deliveroo operates with a ML-driven dispatch engine, which, the company claims, is called 'Frank' and is 'constantly calculating and recalculating the best combination of riders to orders ... rider travel time, food preparation time etc.' (Pudwell 2017). From 2016 onwards, strikes by Deliveroo drivers against the low pay and high insecurity enabled by 'Frank' spread virally from the UK across Europe, reaching as far as Hong Kong (Cant 2017; Zamponi 2018). A related stream of struggles is the rolling and complex series of legal challenges mounted from London to California by drivers for Uber against their algorithmically managed non-worker status (Rosenblat 2018). These have unfolded both in tandem and in contradiction with the legal objections, street blockades and protest suicides of 'regular' taxi drivers fighting the undercutting of their work. Intense as these tumults are, they pale in comparison with the scale of driver strikes, protests, algorithm sabotage and system-gaming raging across the vast operations of Didi, China's Uber equivalent, insurgencies that may prefigure the chaotic future of labour activism under AI-capital (Chen 2018). To the file of worker actions should be added the unexpected North American unionization wave in new digital journalism enterprises such as Gawker, Salon, Huffington Post, Vice Canada and Al Jazeera involving young workers under 'immense pressure to generate 24/7 online and social media content' whose earnings and labour conditions are directly shaped by AI-driven advertising and directed news feeds (Cohen and de Peuter 2018: 115). More diffuse forms of worker organization are to be found in the microwork platforms that, as we have seen, are part of the AI production process. Turker Nation, the online forum of labourers on Amazon's Mechanical Turk, has for years fostered microworker mutual support and knowledge sharing, challenged egregious exploitation by the platform's corporate clients and protested the policies of Amazon itself (Harris 2014; Katz 2017). The more ML and other forms of AI suffuse capital, the more all worker resistance to corporate power will involve *de facto* resistance to the algorithmic forms in which that power is instantiated.

2) Protests against military and paramilitary AI applications. Over the summer of 2018, a remarkable wave of technology-worker resistance to military and militarized policing projects swept through Silicon Valley (Dyer-Witheford and Matviyenko 2019). At Google, workers organized to shut down Project Maven, a Pentagon project that uses ML to improve targeting for drone strikes, and succeeded. Following this, Google's CEO, Sundar Pichai, published a statement of principles on AI development, saying the company would not work on AI weapon or surveillance contracts, although reserving the right to pursue other military projects. The company also withdrew its bid on a $10 billion Pentagon contract, the Joint Enterprise Defense Infrastructure (JEDI) cloud computing project (BBC 2018a). Meanwhile, at Amazon, workers petitioned Bezos to stop selling the corporation's Rekognition facial identification software to US police departments and the Immigration and Customs Enforcement (ICE) agency, notorious for its 'zero-tolerance' enforcement policies. At Microsoft, workers similarly demanded the termination of a $19.4 million cloud computing contract with ICE (Frenkel 2018), and at Salesforce, workers tried to block the company's involvement with Customs and Border Protection (CBP). Alongside these events, and belonging to issues of sexism and discrimination discussed in point five below, we can include the worldwide walkout of Google employees against the corporation's handling of sexual harassment issues. Dissent against the security state is nothing new, but the Silicon Valley revolt was exceptional. It mobilized an elite, high-technology workforce centrally involved in the production of digital weaponry that had for decades been considered immune to serious politicization by virtue of its wealth, status and libertarian mind-set. Such events do not come out of nowhere. In part they indirectly reflect an increasing restlessness among high-tech workers whose wages, good as they are, have not kept pace with the obscene expanding wealth of their bosses. Behind the outbreak, however, lay organizing by groups such as the Tech Workers Coalition, which for years had worked with labour organizations to unionize contracted programming work and patiently campaigned in Silicon Valley and Seattle to develop a critique of corporate power, opposition to racism and sexism, and solidarity between programmers and low-wage support workers (Tech Workers Coalition 2018).

3) Anti-surveillance movements. Since Edward Snowden's revelations, awareness of the scope of surveillance in capitalist liberal democracies has escalated (Gallagher 2018) (though of course such surveillance is no news to populations elsewhere – for example in China). This is an AI issue, not just because the NSA uses ML in its scans, but because surveillance is crucial in the development of AI. As we have described, large data sets are necessary to train ML systems. Not all of these data sets necessarily scrutinize human behaviour, and there are some technical attempts to mitigate the invasiveness of such observation (Knight 2016b; Marr 2018), but AI development is likely to remain significantly reliant on large-scale people watching. In North America, anti-surveillance self-defence manuals and information sheets have proliferated across social movements. 'User's guides for privacy and protest' summarize practices of 'obfuscation – the deliberate addition of ambiguous, confusing, or misleading information to interfere with surveillance and data collection' (Brunton and Nissenbaum 2015). Facial recognition AI, and the possibility of baffling it with 'adversarial images', is a recent hot-spot. There are, however, serious limitations of time, expense and expertise on individualized anti-surveillance techno-measures. Political and legal limits to surveillance have better chances of mass effect. In North America, some of the most important initiatives are from minorities that know only too well how profiling systems at once constitute and control suspect social groups (Dyer-Witheford and Matviyenko 2019): African Americans, opposing the digital tracking both of entire neighbourhoods and Black Lives Matter leaders; First Nations fighting the digital scrutiny of land and pipeline protests; and Muslim Arabs, opposing wholesale databank inventories of their existence. The major political steps in surveillance restriction have, however, come from Europe, where memories of Nazism and Stalinism are still alive. Anti-corporate surveillance movements such as Max Schrem's 'None of Your Business', and governing elites' opposition to national sovereignty infringements give the issue real political weight. The European Union's General Data Protection Regulation (GDPR), which came into force in 2018, limits the data gathering of digital capital, including the great AI developers. Its requirements for active user opt-in (rather than default acceptance) of digital tracking, corporate transparency about the use of personal data, users' right to leave platforms without losing stored data, explanations of automated decisions, and for a 'right to be forgotten' all crimp the use of population data as extractive raw material for the construction of machine intelligence (Meyer 2018; Kaput 2018; Srnicek 2018). The negative response of ML developers and computer scientists, and their critical comparison of Europe's regulations with the corporate freedoms of the United States and the powers of China's surveillance state (K-F. Lee 2018), indicate how big a problem anti-surveillance movements pose to AI-capital, but also the challenges such movements face in an AI-competitive world-market.

4) Social media defection. More diffuse and invisible than anti-surveillance activism is dissent by subtraction. In the wake of the Cambridge Analytica scandal, leaving social media paradoxically became a social media meme, driven by an amalgam of concerns over digital surveillance and manipulation, the psychological effects of techno-addictions, and revulsion at the extreme commodification of online environments. A 'Delete Facebook' campaign sprung up, then subsided. There is, however, evidence that people in the 18 to 24 age range are either curtailing or, occasionally, abandoning social media use, even while their parents become increasingly immersed (Kale 2018). In so far as ML is a big data undertaking, some AI-capital may be impeded by a slow exodus of data-subjects 'logging off' social media.

5) Algorithmic bias busting. We have seen the tendency of AIs to amplify pre-existing sexist and racist discrimination. Businesses owned and dominated by white men, and governance systems administered by the same, are producing AI systems that discriminate against women and minorities because they are trained on data sets that reflect the historical levels of hiring, wealth, clean records and career success enjoyed by white men (Noble 2018). Such discriminatory AI affects those who encounter it on the screens of police, border guards and welfare services with particular severity. That algorithmic bias has even been named is, however, evidence of struggles against it. Groups addressing the issue range from the Algorithmic Justice League, fighting 'the coded gaze' (Buolamwini 2018) at the Massachusetts Institute of Technology, to NGOs like the San Francisco-based, non-profit Human Rights Data Analysis Group (HRDAG), which revealed algorithmic bias in US predictive policing software, to community organizations such as Silicon Valley De-Bug, which is involved in immigrant and prisoner rights and economic justice in the San Jose area, or Philadelphia's Media Mobilizing Project, which is organizing low-wage workers and poor communities

(Wykstra 2018). As correcting algorithmic bias increases the accuracy of AI predictions, it can, in theory, be serviceable to capital to improve its niche targeting of commodities and hiring of talented labour-power, even if this requires acknowledging embarrassing and expensive displays of sexism, racism and homophobia: these, at least, are the terms progressive neoliberalism sets for such initiatives.[2] However, in so far as algorithmic bias discriminates against people who are both poor and female, queer or racialized – hence the most vulnerable of the vulnerable – algorithmic justice movements hamper AI-capital's disciplining and discarding of its so-called surplus populations, challenging the 'matrix of domination' that has informed its development (Costanza-Chock 2018).

6) Digital city disturbances. As giant AI developers impress an ever stronger footprint on the urban landscape with their headquarters, campuses and experimental sites, social conflicts explode in the city-incubators of AI. Nowhere is this more evident than around Silicon Valley, where billionaires in luxury mansions, millionaires working in high-tech campuses, low-paid service workers and a large homeless population coexist in constant tension. As software production expanded north from the Valley of San Francisco and the Bay area, so too did these social fault-lines. In 2013–14, the private buses that transport high-tech workers to the 'Googleplex' in Mountain View became a flashpoint for protests against gentrification, eviction, displacement and congestion (Goode and Miller 2013). Demonstrators blockaded the bus routes. Signs read 'Gentrification and Eviction Technologies: Integrated Displacement and Cultural Erasure' and 'Fuck Off Google'. Leaflets accused Google of 'building an unconscionable world of surveillance, control and automation' (Streitfeld and Wollan 2014). The protests flared up, went out, and then re-appeared in 2018. 'Techsploitation is Toxic' militants blocked a dozen Google buses by piling in front of them the electric scooters used by an Uber-style company, and denounced plans to clear the streets of homeless encampments with slogans such as 'Sweep tech not tents' and 'They call it "Disruption." We call it displacement' (Streitfeld 2018). These urban disturbances also include contestation of AI-based 'smart cities'. In 2017, Sidewalks Labs, a subsidiary of Google's holding company, Alphabet, negotiated with the city of Toronto a 'public-private' partnership for the development of the waterfront neighbourhood Quayside. Under this pact, the details of which were not disclosed, Quayside would become a showcase for AI-saturated urban innovations; streets designed for autonomous vehicles, robot garbage collection, ubiquitous sensors and security cameras, and a centralized information management system reading residents' ID from library and health cards. Canada's Prime Minister Justin Trudeau praised the project as a harbinger of clean, green urbanity. Eric Schmidt, Google's former CEO, explained that the idea came from Google's founders getting excited thinking of 'all the things you could do if someone would just give us a city and put us in charge' (Kofman 2018). Apparently unmentioned in the initial agreement were the ownership and terms of collection of digital data to be gathered in Quayside. This issue remained obscure until forced into the public eye by a campaign led by TechReset Canada (Barth 2018), a group opposed to unrestrained digital commodification that criticized the opacity of the deal, insisted the contract be made public, and demanded provisions to ensure that 'the data and data infrastructure of this project are the property of the City of Toronto and its residents'. This mobilized a coalition whose participants ranged from housing activists to disgruntled city planners; their cause won support even from leading Canadian high-tech entrepreneurs, objecting to intellectual property surrender to foreign corporations. Prominent members of the board overseeing the Quayside agreement resigned. As of early 2019, the issue remains unresolved. Sidewalk Labs may concede to objections, but also dodge the protest, by agreeing to collect only aggregated, anonymized data – a ploy that, while it addresses the limited issue of personal privacy, leaves untouched the larger problems of corporate command over population-level information and urban design (Greenfield 2013).

7) Anti-corporate techlash. Many of these concerns have converged in a nebulous 'techlash' against the power of the large digital corporations. Facebook, Cambridge Analytica and the Russian cyber-war scandals in the 2016 US election catalysed a revival of long-dormant discussions about concentration of ownership in digital oligopolies (Foer 2017; Mosco 2017). Even within a US high-technology community famously opposed to state regulation, voices expressed scepticism as to the capacities of Google and Facebook to police themselves and called for enforced transparency and even financial liability for the negative externalities of their algorithmic processes. Other criticisms went deeper – asking, for example, about the degree to which capital's media corporations are deeply and basically dependent on the 'fake news' of advertising and on highly engineered attention manipulation. In Europe, litigation was launched against the giants of digital capitalism over violations of public regulation and labour law, tax evasion and anticompetitive abuse of monopolistic powers. Calls for the disaggregation and regulation of some of the largest AI-developing corporations appeared in the platforms of Bernie Sanders and Jeremy Corbyn (and even, erratically, in the ruminations of right-wing authoritarian populists such as Trump). Antitrust politics are by no means the same as anti-capitalism, but one effect of the 'mainstreaming' of such concerns has been to open up a space for discussion of more radical propositions for a 'socialization of the data banks' (Morozov 2015). These extend beyond the regulation or breakup of oligopolies to the establishment of 'commons' institutions separate from both state and market, and the formation of public computing utilities kept at arm's length from governmental and corporate power. We return to these issues in the Conclusion. By 2018 even business commentators were speaking of changed attitudes towards digital capital as one of an array of factors, alongside gadget-saturated markets and international trade wars, that were giving high-tech a 'hammering' on that infallible index of human values, the stock market (Rajan 2018).

CONCLUSION: TOWARDS THE EVENT HORIZON

Some working-class recomposition *is* occurring in digital capital, with the powerful presence of AI as an invisible attractor around which various rebellions and dissents array themselves. We do not want to overstate this process. Anti-corporate techlash can be politically shallow; smart city opponents are still marginal; algorithmic justice both hinders and helps capital; defections from social media are slow and partial; anti-surveillance policies could be rolled back by competitive pressure; despite the success of the Project Maven revolt, other victories against militarized AI are scant; above all, over the uptick in strikes against AI-capital hangs the shadow that resistance may spur further automation. While we have suggested that, regardless of job-loss issues, AI-capital poses many other threats to the well-being of its populations, it is also the case that the future of struggles within it depends heavily on how deeply the fourth industrial revolution undermines the basis of wage-labour.

As we have seen, early predictions of an imminent AI-caused employment implosion have been significantly watered-down by more recent studies, and have lost their edge as the post-recession US economy returned to what is officially regarded as full employment. Yet some observers suggest even limited, sectoral job loss to automation could figure as largely in upcoming US elections as the jobs-gone-abroad issue did in 2016 (F. Levy 2018). Longer term, estimates of AI employment effects continue to vary wildly, and even some Business-as-Usual theorists have changed their models to admit the possibility that capital's job-creating and job-destroying functions might go seriously out of sync (Acemoglu and Resteropo 2018). Some maintain quite catastrophic positions; the widely read treatise on AI by Kai-Fu Lee (2018), former head of Google China, insists on a loss of up to 40 per cent of US jobs by 2030, and similar numbers for China – in other words, numbers of world-historical turbulence. Such runaway AI, massively increasing the reserve army of the unemployed, would further weaken unions and other worker organizations, but the extremity of immiseration could generate new forms of proletarian revolt (even, perhaps, as bloody and determined as those of *Westworld*'s androids). In such speculations, we arrive at the event-horizon foreseen by the 'Fragment on Machines', where capital's own machinic drive explodes it from within: terminally decomposing its working class, capital decomposes itself. It is the apparently irrefutable logic of this futurist proposition that perhaps gives leftists a certain last-ditch confidence, complacency or even enthusiasm as they face AI-capitalism. 'Probably', one says, 'the tech won't really work – and, anyway, if it does, it only perfects the system's self-destruction.' But how warranted is this assurance?

3

Perfect Machines, Inhuman Labour

Narrow AI is a special case of capitalist machinery that challenges Marxist thought from numerous directions. This chapter, however, turns to artificial general intelligence (AGI), or AI with capacities for reasoning with general knowledge and doing a variety of tasks in diverse and unfamiliar domains (Gubrud 1997; Wang and Goertzel 2007: 5; Goertzel 2014: 2; Baum 2017: 3). AGIs are thus machines with a flexibility or generality of intelligence that is similar in scope, though not necessarily in functioning, to that of humans. As Pei Wang and Ben Goertzel explain, an AGI could be used 'in situations where ready-made solutions are not available, due to the dynamic nature of the environment or the insufficiency of knowledge about the problem … what we expect from an AGI system are not optimal solutions … but flexibility [and] creativity' (2007: 5).

These references to flexibility and creativity bear a striking resemblance to Marx's concept of labour. Marx, however, presupposed that 'labour … is an exclusively human characteristic' (1990: 283–4) and that the capacity to labour exists only 'in the physical form, the living personality, of a human being' (1990: 270). Against this anthropocentrism, we argue that there is an isomorphism between the theoretical notion of AGI and Marx's concept of labour and labour-power; AGI, therefore, profoundly challenges Marx's labour theory of value; in particular the axioms that only human beings can labour and create value, and that machines categorically cannot. AGI is, in other words, such a special case of capitalist machinery that its status as machinery must be questioned. We argue that AGI could not only potentially labour, but under certain social conditions also create value. The potential ramifications of AGI for the capitalist mode of production and humankind are drastic. Were AGI to be created, many believe it would be the 'technological singularity', that is, an outcome of 'exponential technological progress' resulting in 'such dramatic change that human affairs as we understand them today c[o]me to an end' (Shanahan 2015: xv).[1] While the further research and development of narrow AI will likely see the pool of people superfluous to capital accumulation grow, the development of AGI would go beyond this already miserable vision as discussed by Marx to an even worse one, that of *Homo sapiens* becoming a *surplus species*. AGI thus suggests the possibility of a capitalism without human beings.[2]

To make this argument, we discuss why Marx assumed both labour and value creation to be inherently human, zeroing in on the sharp distinction he makes between humans and animals. We consider why he grounds this distinction in various capabilities like creativity and imagination, and how some AI projects, as of 2019, are trying to automate such capabilities. While the previous chapters have engaged in some necessary speculation, this chapter envisages a distant future that is more science fiction than science fact. Thus, before we delve into the specifics of Marx's value theory and his arguments about human uniqueness, we first justify why we are devoting a chapter to a technology that has yet to, and might never, see the light of day.

ARTIFICIAL GENERAL INTELLIGENCE

No scientific or philosophical consensus exists on the definition of intelligence, but most perspectives agree that true intelligence must be to some degree flexible or general (Kaplan 2016; Lake et al. 2017: 9–19). Existing ML and GOFAI are distinguished from human intelligence by their narrowness and inflexibility. Research is, however, underway to produce AI which can be generally applied – AGI. The term AGI was first introduced by Mark Avrum Gubrud (1997) to describe 'AI systems that rival or surpass the human brain in complexity and speed, that can acquire, manipulate and reason with general knowledge, and that are usable in essentially any phase of industrial or military operations where a human intelligence would otherwise be needed.' The idea of AGI, however, actually dates back to the original goal of the 1956 Dartmouth workshop (Wang and Goertzel 2007: 6). But the goal of emulating human intelligence was quickly abandoned in favour of more manageable practical problems that could be solved by narrow AI, and until the early 2000s AGI was considered so unrealistic that people working on it were seen as crackpots (Pennachin and Goertzel 2007: 1; Wang and Goertzel 2007: 4). While there were research projects that were focused on the goal of building AGI, such as the Soar cognitive architecture that John Laird, Allen Newell and Paul Rosenbloom started in 1983 (Laird

2012) and the Cyc project initiated by Douglas Lenat in 1984 (Lenat and Brown 1984), it was not until the fiftieth anniversary of the Dartmouth workshop and the rise of ML in the early 2000s that the original goal of AI, reincarnated as AGI, has again become acceptable and seen as realistic (Goertzel 2007: 1162).

Despite AGI sharing the original goal of AI, what the former refers to differs significantly from what AI has come to mean as it has been applied commercially in the 2010s. Whereas narrow AI focuses on solving narrowly constrained problems, the goal of AGI research and design is to create systems 'with sufficiently broad (e.g. human-level) scope and strong generalization capability' so that it is able to transfer knowledge 'from one problem or context to others' (Goertzel 2014: 2–3).

A term that appears to be near-synonymous with AGI is 'human-level machine intelligence' (HLMI).[3] What is interesting about the definition of this term is that it is grounded in labour.[4] One definition refers to HLMI as AI that 'can carry out most human professions at least as well as a typical human' (Müller and Bostrom 2016: 558), and another posits that it 'is achieved when unaided machines can accomplish every task better and more cheaply than human workers' (Grace et al. 2017: 1). A machine that is capable of doing many different types of jobs would almost by definition have to be generally intelligent if it were to master jobs that it had not been taught to do. While HLMI would appear to be a perfect fit to discuss the possibility of increasingly intelligent machines that eventually become capable of labour, we reject this term because of its anthropocentric notion that a machine can be defined as intelligent only if it is human-like. As Benjamin Bratton (2015) argues, 'human intelligence simply can't exhaust the possibilities' of all forms of intelligence because we 'would do better to presume that in our universe, "thinking" is much more diverse, even alien, than our own particular case'. AGI is therefore a preferable term to HLMI because it 'stress[es] the "general" nature of the desired capabilities of the system' (Wang and Goertzel 2007: 1). An AGI could possess intelligence equally general to that of a human being without mimicking human cognition. This difference of cognition is already readily clear with how narrow AI does things compared to humans. For example, 'AI chess programs will use brute force searches in some instances in which humans use intuition, yet the AI can still perform at or beyond the human level' (Baum 2017: 8). In addition, AGI is a preferable term because, as Ben Goertzel argues: (1) humans are not necessarily that smart, hence, 'human-level' may actually be limiting the scope of what AGI is; and (2) if it is possible to conceive of and thus develop 'an AGI system with different strengths and weaknesses than humans, but still with the power to solve complex problems across a variety of domains and transfer knowledge flexibly between these domains', then it becomes tough to define whether the system is human-level (2007: 1163). The arguments made by Bratton and Goertzel are redoubled by those who suggest that what is really at stake in AI is the dissociation of intelligence and consciousness (Shanahan 2015; Harari 2016), that is to say, the possibility that a machine could behave intelligently without possessing a self-awareness similar to that experienced by humans.

Let us reiterate: as of 2019, AGI does not exist. Some doubt that such machine intelligence is even possible (Dreyfus 1972; Braga and Logan 2017; Nadin 2018). Within the AI community, however, many consider AGI to be possible at least in theory and some argue that it 'is merely a very difficult engineering problem' (Pennachin and Goertzel 2007: 1). At the very least, AGI is considered sufficiently plausible to attract investments for research and business ventures.

Within the larger AI community, AGI researchers form a very small minority. The narrow AI industry dwarfs the AGI one in terms of the number and size of corporations and research institutions, and projects and funding. Baum's 2017 survey of active AGI projects gives an overview of the political economy of AGI as an industry. This survey identifies a total of 45 active AGI projects[5] in 30 countries worldwide, with most of them being 'based in major corporations and academic institutions' and a smaller number based in public corporations, nonprofits or governmental institutions.[6] While 'some ... are large and heavily funded', most of them are small to medium in scale (Baum 2017: 2). AGI research is characterized by having a large number of open-source projects, with a total of 25 projects having made their source code available online. While only nine of the AGI projects have identifiable military connections, only four were identified as clearly having no military connections with the rest being unclear. Most AGI projects are based in the US or within its sphere of influence, and the only ones that are not are based in Russia and China (Baum 2017: 2).

The largest AGI projects are Alphabet-Google's DeepMind, the Elon Musk-backed Open AI, and the Human Brain Project, while other notable projects include Vicarious FPC, the Microsoft acquisition Maluuba, Open Cog, Uber AI, and Nnaisense. According to Baum, there are two trends among the for-profit corporations: either to support 'long-term AGI R&D in a quasi-academic fashion, with limited regard for short-term profit or even any profit at all', or to undertake 'AGI projects delivering short-term profits for corporations while working towards long-term AGI goals' (2017: 19) The latter trend is apparent in corporations like Microsoft, Google and Uber; Baum argues that if 'this synergy between short-term profits and long-term AGI R&D proves robust, it could fuel an explosion of AGI R&D similar to what is already seen for deep learning' (2017: 19). Based on Baum's survey, it appears that the current state of AGI research is still primarily, as Pei Wang and Goertzel argued a decade prior, 'producing publications and preliminary results' (2007: 3).

How far off we are from inventing AGI is uncertain, but many working in the field of AI view it as inevitable. Ethem Alpaydin argues that, based on what ML systems can do today, it 'will not be surprising if this type of learned intelligence reaches the level of human intelligence some time before this century is over' (2016: xii). There are, however, more optimistic estimates. In a survey asking AI experts when they expected HLMI to be developed, the median estimate was a 10 per cent probability by 2022, 50 per cent probability by 2040, and 90 per cent probability by 2075 (Müller and Bostrom 2016). Another survey asking ML experts when AI will exceed human performance yielded similar results: 10 per cent chance of occurring within nine years, 50 per cent within 45 years, and 75 per cent within 100 years (Grace et al. 2018). This survey also asked granular questions about when AI will outperform humans in a particular skill, giving estimates for many different tasks, including translating languages by 2024, driving a truck by 2027, working retail by 2031, and writing a best-selling book by 2049. The ML experts believed that there is a 30 per cent chance of AI outperforming humans in all of these tasks in 45 years, and that all human jobs would be automated in 120 years.[7]

AGI is actively pursued by corporate, governmental and nongovernmental interests. The actual invention of such machines may be one of the bleakest possible futures of capitalism for humans because, as we now critically discuss, such a machine could labour and therefore potentially also create value. Although being able to perform labour is a necessary condition for creating value, it is not a sufficient one. We therefore proceed analytically and consider the possibility of AGI labouring separately from whether it could create value.

HUMAN LABOUR AND LABOUR-POWER

In *Capital*, Marx presupposed labour as 'an exclusively human characteristic' and repeatedly qualified this central concept as 'human' and 'living' or even 'living human' (1990: 283–4). His qualification of labour in this way can be explained as rhetoric against the bourgeois political economists who fetishistically argued that land and capital, in addition to labour, were sources of value.[8] But Marx's assumption that labour is inherently human is more than rhetoric. While Marx primarily discussed labour with reference to the social relations in which it occurs, the touchstone for his concept of labour was, for better or for worse, the human being. That labour is human is one of Marx's axioms.

When Marx discussed labour as such, i.e. without reference to the society in which it occurs, he sharply distinguished it from the productive activity of animals. In Marx's theoretical framework, when animals are employed in a capitalist production process, unless as raw-material inputs, they are functionally equivalent to machines, reduced to the status of fixed capital.[9] Given this functional equivalence, the arguments Marx advanced in the *1844 Manuscripts* and *Capital* for why labour is qualitatively different in kind from the productive activities of animals are salient for understanding both why machines cannot labour – and why, in theory, AGI could labour. Marx's arguments for why animals cannot labour are especially interesting because they describe animals in terms quite similar to the domain-specific behaviours of narrow AI. By extension, this suggests an isomorphism between the concept of AGI and Marx's concepts of labour and labour-power.

In the seventh chapter of *Capital*, Marx argues in transhistorical terms that labour is 'a process between man [*sic*!] and nature, a process by which man through his own actions, mediates, regulates and controls the metabolism between himself and nature', and through this metabolism the human species 'appropriate[s] the materials of nature in a form adapted to [its] own needs' and 'acts upon external nature and changes it' (1990: 283). The statement makes it appear as if Marx is arguing that humans are the only beings that act on, use and alter their lived environment. In the *1844 Manuscripts*, however, he recognizes that all animals are 'sensuously acting, objectifying subjects' that build their own versions of the world and reproduce their habitats, albeit in a species-specific manner (Fracchia 2017). Throughout the *1844 Manuscripts*, Marx recognizes that humans and animals have a lot in common because all living beings must engage in a 'continuous interchange with nature' in order to survive (Mulhall 1998: 10). In addition to producing species-specific habitats, humans share with animals the 'animal functions' of eating, drinking and procreating, etc. (Marx 2007: 75). But despite the similarities, the specific way in which humans engage with their external environment differs significantly from that of animals.

Marx explains this difference with the concept of life-activity and argues that the 'whole character of a species ... is contained in the character of its life-activity'. Whereas an animal 'is immediately identical with its life-activity ... does not distinguish itself from it [and] is *its life-activity*', humans are distinguished by having 'conscious life-activity' (2007: 75). One of the dividing lines Marx thus draws between humans and animals is that the former are conscious of what they do, while the latter are not. He reiterates this argument in *Capital*, stating that the conscious purpose behind what humans do 'determines the mode of [their] activity' (1990: 284). In turn, this consciousness gives humans 'free play of [their] own physical and mental powers' (1990: 284); they are therefore 'free' in their activity (2007: 75).

Why does consciousness grant human labour-power freedom? What does Marx mean by 'free'? The implication is, of course, that in their activity animals are unfree. But why? This difference in relative freedom can be better explained by turning to two passages where Marx underscores the difference between humans and animals. Importantly, these passages also identify other mental capacities Marx argued were uniquely human. In the *1844 Manuscripts*, Marx writes:

> Admittedly animals also produce. They build themselves nests, dwellings, like the bees, beavers, ants, etc. But an animal only produces what it immediately needs for itself and its young. It produces one-sidedly whilst man produces universally. It produces under the dominion of immediate physical need, while man produces even when he is free from physical need and only truly produces in freedom therefrom. An animal's product belongs immediately to its physical body, while man freely confronts his product. An animal forms things in accordance with the standard and need of the species to which it belongs, whilst man knows how to produce in accordance with the standard of every species and knows how to apply everywhere the inherent standard to the object. Man therefore also forms things in accordance with the laws of beauty. (Marx 2007: 75–6)

This should be read alongside the (in)famous bees and architect passage from *Capital*:

> A spider conducts operations which resemble those of the weaver and a bee would put many a human architect to shame by construction of its honeycomb cells. But what distinguishes the worst architect from the best of bees is that the architect builds the cell in his mind [*Kopf*] before he constructs it in wax. At the end of every labour process, a result emerges which has already been conceived by the worker at the beginning, hence already existed ideally [*ideell*]. (Marx 1990: 284)[10]

Here, Marx homes in on consciousness and cognition (e.g. creativity, learning and imagination) as uniquely human and opposes these to the instincts of animals, which determine their entire life-activity. Both human and animal production is hardwired to the survival instinct, but instinct also determines what and how, and the times and places animals produce and consume. The form and content of an animal's existence is thus determined by what species it is because all animals are identical with their life-activity (Mulhall 1998: 12; Fracchia 2017). For Marx, animal activity is thus 'unfree' because it is instinctual.

Humans thus differ from both animals and algorithms because they have 'cast off' the 'instinctive forms' of their life-activity, i.e. labour (Marx 1990: 283). The anthropocentrism of Marx's concept of labour can be understood as based on an assumption Marx made about the evolution of humans and of other animals: the former evolved a consciousness and a series of other mental capabilities while the latter did not. Therefore, humans can make their life-activity and labour 'an object of will and consciousness', meaning that they can reason about what, how, when and why they produce, and can imagine and deliberate on what they *could* do differently (Marx 2007: 75; Mulhall 1998: 13). It is on the basis of such choices humans can make in production that Marx makes the argument that humans have 'free play' of their mental and physical capabilities.

This free play also enables humans to produce 'universally', i.e. across multiple domains, as opposed to the one-sided, domain-specific production of animals. In other words, humans have a broader range of productive activities than other animals, a range that includes producing 'in accordance with the standard of every species'. In addition, human production is not limited to immediate physical need because whatever natural limits were once imposed by instinct are overcome and 'artificially expanded' (Fracchia 2017). In principle humans can, according to Marx, put their minds and hands to almost anything, including imitating nature and learning from what other animals do and make, such as bees' honeycomb cells, beavers' dams and spiders' webs, and 'produce as artifacts that which "immediately belongs to the physical body" of an animal (clothes as artificial skin) and can learn and make things to do what other species do "instinctually" (winged vehicles to fly)' (Fracchia 2017).

But why can humans do all of this with their labour? Marx's argument that humans are able to produce universally should be understood in terms of them being able to, first, learn from what they have observed in nature about what animals do, and from what they have succeeded or failed at in the past, and then, second, through their creativity or imagination, apply this new knowledge to a different domain of production, or even invent a new one. In turn, this ability is dependent on the various mental capabilities or phenomena that Marx understands as uniquely human. In the two passages quoted above, Marx explicitly or implicitly grounds human labour in: aesthetic appreciation; learning and understanding (from what other animals produce); a conscious mind; forming ideas; imagining and conceiving plans; and, in general, being creative and adaptive when solving the problem(s) of production. In the aggregate, these mental capabilities can be understood as general intelligence, which makes humans flexible in what they can do and adaptable to new and different environments, objects of labour, and other conditions of production.

HUMANS, ANIMALS AND MACHINES TODAY

There are many problems with Marx's assumptions and arguments about both humans and animals, as well as the work of architects. At best they were based on nineteenth-century evidence from the nascent science of biology, at worst on his imagination and humanist beliefs. Can the sharp distinctions he draws between humans and animals be sustained in the light of contemporary ethology? Why cannot animals labour? Can Marx's assertion that labour is inherently human be sustained? The debate about whether animals can labour and/or produce value is decades-old (see e.g. Benton 1988, 1993, 2003; Elster 1985; Wilde 2000; Perlo 2002; Drake 2015; Johnson 2017; Kallis and Swyngedouw 2018; Morton 2017), and Marx's sharp distinction between the freedom of human labour and the unfree instinctual-algorithmic behaviour of animals has been critiqued using evidence from ethology, i.e. the study of animal behaviour and cognition. Jon Elster (1985) and Ted Benton (1988, 1993), for example, argue that animal production is far from one-sided, being characterized not only by a high degree of variability but also by considerable adaptability, especially when they are confronted with new environmental conditions (such as urban settings). Similarly, Paul Cockshott and Karen Renaud challenge the sharp distinction Marx draws between humans and animals on the basis of planning or having a vision (i.e. imagining) before they do something, noting how wolf packs divide tasks while hunting, how various mammals use episodic memory to plan, and how experiments have shown that spiders form plans while web-weaving (2016: 21–2).

Animal behaviour is more general than Marx believed and so is their cognition. Indeed, researchers of cognitive evolution no longer posit that animal behaviour can be fitted into a hierarchy of complexity, but instead focus on identifying and describing animals' 'mental organs' or modules that each have a domain-specific function and evolved to only accept a 'certain kind of input' (e.g. optical, aural or olfactory) in relation to a given ecological problem (Vitti 2013: 394). While initially specialized, 'modular cognition', due to increasing encephalization, evolved into 'flexible and domain-general' cognition (Vitti 2013: 395). What has driven this evolution may be 'omnivorous diets, variable habitat, long developmental periods, extended longevity, and pervasive social interactivity' (Vitti 2013: 396). While most animals, the honeybee included, 'specialize in the skills that their ecological niche requires', some species, with high encephalization levels (mammals, birds and cephalopods), are considered 'generalists' (Vitti 2013: 395).

Timothy Morton (2017: 56) takes a more fundamental issue with Marx's argument that the labour of human architects is different from that of bees. He contends that it is impossible to demonstrate that either humans or bees have the capacity of imagination, or that humans are not executing algorithms when they plan. The assumptions Marx makes to argue that human productive activity is uniquely distinct from that of animals is, Morton argues, 'based on a metaphysical assertion about humans' (2017: 56).

What these critiques make clear is that the distinction between humans and animals is more blurred than Marx imagined. We argue that the distinctions between humans and intelligent machines are also becoming increasingly blurred. Indeed, many of the examples of the complex behaviour and inner lives of mammals and insects that are presented against Marx's insistence that labour is inherently human have analogues in the functioning of ML systems. If Marx's conceptualization of labour as human can be challenged with examples of animals doing things previously thought uniquely human, the same holds for machines, especially those that exhibit behaviours that appear to be productive, creative, imaginative and even adaptive. Some ML systems even appear to envision plans or concepts in their neural architectures that are then respectively carried out or recognized in new corresponding examples the system has never before encountered.

A striking example of machine creativity is DeepMind's AlphaGo, which first beat European Go champion Fan Hui in October 2015 and then Go world champion Lee Sedol 4–1 in a match in March 2016. The ancient board game of Go has long been a milestone in AI research due to it being one of the most complex games humans have designed. Chess-playing AIs can simply use brute force and search all potential moves until the winning ones are found, but this is not possible in Go since there are more possible configurations of the board, and thus potential moves, than there are atoms in the universe. Until AlphaGo proved otherwise, the game was seen as requiring human-level pattern recognition and something akin to intuition.[11] To learn how to play Go, AlphaGo relied on a combination of deep neural networks and Monte Carlo tree search. Its architecture consists of three convolutional neural networks (CNN) that were divided into one 'value network' and two 'policy networks'.[12] The CNNs were trained on image data that represented individual moves ('states') from 160,000 games recorded from top Go players (Silver et al. 2016). The value network learned how to play the game and to estimate the probability of winning given any current state of the game, while the policy networks learned what move to select given the current state of the game. In turn, the two policy networks' decision-making was reinforced by having them play each other 30 million times and keeping track of the best outcome of the games (Silver et al. 2016).

AlphaGo's creativity or intuition was evident in the 37th move of the second game against Sedol. Although this move was later described as 'creative', 'unique' and 'beautiful', as well as 'inhuman' by professional and champion Go players (including Sedol), it at first took everyone by surprise (Metz 2016). After AlphaGo made this move, Sedol took an unusually long time to respond. Commentators noted how strange it was, some thinking that the neural network had made a mistake because a human would never make such a move. But the move was deliberate and based on what AlphaGo's policy networks had learned. The engineers at DeepMind could not have predicted this move and could only determine AlphaGo's reasoning after it had made its move: having calculated the probability of a human making this particular move at that particular state of the game at 0.010 per cent, it nevertheless made it (Metz 2016). AlphaGo arguably made a plan and executed it. Not only did AlphaGo learn how humans play, but it went beyond the constraints that generations of human Go players had placed on how to play the game, making seemingly counterproductive and highly novel moves, and playing the game in styles that no human would (Metz 2016). AlphaGo's creativity comes from moves like this and from how it learned to play Go in a completely new and inhuman style; if a human had done any of this, she would likely have been recognized as original, creative and highly intelligent.

Despite this victory, we note that 'Go' in Katja Grace et al.'s (2018) list of AGI milestones has not been passed. This milestone is specifically listed as 'Go (same training as human)' (Grace et al. 2018: 747). While these shifting goalposts can be explained in terms of the AI effect, AlphaGo's victory, while impressive, is less so when compared to how a human learns how to play the game. As Brenden M. Lake et al. (2017) point out, a human being can learn Go after playing just a few times, but for a ML system to do the same, it requires data from hundreds of thousands to millions of games. AlphaGo was trained on over 30 million games of Go, whereas Lee Sedol, has likely played only 50,000 games in his lifetime (Lake et al. 2017: 23). This difference emphasizes the major drawback of machine learning: the excessive amount of data it needs to do just one thing. On the other hand, human beings are able to 'build rich causal models, make strong generalizations, and construct powerful abstractions', all based on extremely limited data (Tenenbaum et al. 2011: 1279). Indeed, the capacity to generalize based on just a few examples may be at the core of general intelligence.

Some cognitive psychologists suggest that human beings generalize from limited data by using their imagination; a path towards AGI may thus lie in overcoming the problem of data gluttony by equipping systems with artificial imagination (Lapovsky 2014; Johnston 2008: 408; Ritter et al. 2017). Imagination is yet another factor that Marx argued marked labour as uniquely human: the human architect builds something in the mind before construction begins in the real world.

The start-up Vicarious PFC has built a system for solving the CAPTCHA codes ('Completely Automated Public Turing Test to tell Computers and Humans Apart') that are specifically designed to distinguish humans from online robots. It rests on giving AI what Vicarious claims is an 'artificial imagination' (George et al. 2017). According to this company, imagination concerns the 'ability to picture what the information [the AI has] learned should look like in different scenarios' (Knight 2016a). More precisely, the company ties such an imagination to the inductive biases and constraints in the visual cortex and neocortex, which cognitive psychology and neuroscience have shown are connected to the learning efficiency and generalization ability of the human brain (Tenenbaum et al. 2011; George et al. 2017: 1).[13]

Inductive biases – also referred to as learning biases – have been defined as 'factors that lead a learner to favour one hypothesis over another that are independent of the observed data. When two hypotheses fit the data equally well, inductive biases are the only basis for deciding between them' (Griffiths et al. 2010: 1). In other words, inductive biases are heuristic processes which the human brain has evolved to deal with the problems of perceiving with and learning from limited data and extrapolating learned experiences to novel contexts. Cognitive and developmental psychologists have proposed a number of inductive biases that human children exhibit, some of which have been emulated in AI projects to identify words and objects (George et al. 2017; Ritter et al. 2017). One inductive bias is the recognition of shapes, which humans prefer over colour, texture or size to categorize objects and words. A predisposition to recognize shapes, and thus identify useful objects, has obvious evolutionary benefits. Shape bias can be understood as an innate interpretive tool that children use when learning new words.

It was through simulating such inductive biases that Vicarious equipped their CAPTCHA-solving AI with an artificial imagination. CAPTCHA codes consist of letters crowded together in distorted combinations and often have added clutter in order to confuse ML algorithms, such as classifiers. These algorithms work well if individual characters can be segmented, but doing this 'requires an understanding of the characters, each of which might be rendered in a combinatorial number of ways' (George et al. 2017: 1). Achieving such an understanding to solve even just one specific style of CAPTCHA typically requires a system to be trained on millions of labelled examples of letters in various combinations of shape and orientation. If not, the system has difficulty in telling what the letter is, where it ends, and where the next one starts. Humans, of course, can solve a new style of CAPTCHA without any explicit training because of the inductive biases of the visual cortex. The feedback connections in the visual cortex of mammalian brains enable mammals to identify a figure from the background (i.e. figure-ground perception) and to isolate 'the contours of an object even when partially transparent objects occupy the same spatial locations' (George et al. 2017: 1).

What Vicarious has referred to (or slyly marketed) as imagination is more generally referred to in the AI community as 'one-shot learning', which is a human capability enabled, at least in part, by inductive biases. One-shot learning refers to a human child or ML system being shown a single labelled example and from this being able to learn how to identify other examples of the same category of objects, be it cats, letters or whole words, by relying on various innate inductive biases. For example, the typical human two year old can learn and start to use a new word like 'cat', and, after seeing just one example, can recognize a cat from various angles, cats of different colours and sizes and in various poses, and so on. For a ML system to recognize cats in photos, it has to be shown millions of examples of cats from different angles, of different shapes, sizes, materials, colours, and so on (Tenenbaum et al. 2011: 1279; Lake et al. 2017: 22). Similarly, after playing just one or a couple of games of Go, almost any human would have learned how to play the game and could also generalize this knowledge to play one of the many variants of Go with little difficulty, but for AlphaGo to do the same would 'require significant reprogramming and retraining' (Lake et al. 2017: 23). Researchers from DeepMind have found that the one-shot learning models of matching neural networks trained on ImageNet also exhibit biases towards shape, although the magnitude of the bias depends on the particular neural architecture (Ritter et al. 2017).[14] Through this inductive bias the neural network learns how to learn categorization or how to be a classifier, i.e. an algorithm that implements classification.

Vicarious simulated inductive biases with a 'structured probabilistic generative model framework' on what they call a Recursive Cortical Network (RCN). Generative models produce representations or abstractions from observed or learned phenomena based on a probabilistic model. To explain using Marx's architect, if she wanted to build a honeycomb-like structure, this architect would have observed honeycombs in a bee's nest; the memory of these honeycombs would be the architect's generative model in her mind, from which she can generate (i.e. imagine) many different honeycomb architectural drawings. In AI, such generative models are first created based on the patterns it has recognized in the data and are then used to predict probabilities from this model, such as producing new variations of the data it has been trained on. In the case of Vicarious' RNC, its inductive bias enabled it to learn the contours and surfaces of alphabetic letters based on little training data and then, in turn, break CAPTCHAs by generatively segmenting characters (George et al. 2017). This generation can be understood as an artificial imagination because segmenting letters in this way is akin to imagining what they would look like in different shapes, orientations and positions or without a particular background or distortion. In this case, imagination appears to be connected not only to an act of creation, but also to adapting to different situations or noisy contexts. For Vicarious, the end goal of equipping AI with an imaginative capacity is to enable it to transfer learning to a completely different domain. Importantly, this transfer can be understood as adaptability, and this word is what AI researchers use to explain how AI can become generally intelligent and/or more human-like (Alpaydin 2016: ix, xii, 17, 24; Kaplan 2016: 49–50).

WHAT LABOURS?

The differences between humans and animals are not as sharp as Marx believed, and the distinctions between human and machines are becoming blurrier the more sophisticated ML systems become. However, do any of these examples of animal and machine production or seemingly imaginative, creative and adaptable behaviour actually challenge Marx's assumption that labour is an exclusively human characteristic? Proposing that Marx's anthropological conceptualization of labour and value creation is void by pointing out that whatever he refers to as uniquely human is also found in the wider animal kingdom is, as Lawrence Wilde argues, no more than a rhetorical strategy whereby a scientific source is used to 'blur the distinction in productive techniques without genuinely getting to grips with Marx's argument' (2000: 44). It would also be a mistake to equate the narrow capabilities and particular functioning of even the most sophisticated deep-learning system with labour. While Marx's anthropocentrism must be challenged, merely listing a set of behaviours or productive activities does not overturn his argument. As long as AI systems can only perform one or even a few narrowly defined tasks, it does not matter how creative or imaginative they appear to be, their behaviour is not labour.

The reason Marx gives for why human labour is different from the activity of both animals and machines is that it is highly flexible, adaptable – in short, general. And while many animals are more adaptable than Marx thought, and have domain-general intelligence, human intelligence is yet more general, which is what gives this particular species a wider qualitative range of productive behaviours than any other. Although Cockshott and Renaud (2016) seek to blur distinctions between what humans and animals can do, they nevertheless recognize that the qualitative range of what humans can do is *wider* than those of all other animals and current machines. Noting how easily Watt's steam engine replaced horses as a source of power in production, and that humans are not as easily replaceable (although they recognize that machines do take their place), they argue that 'there is more to work than muscular energy' and, hence, that human beings 'can be set to almost any labouring task ... they are adaptable ... it is this adaptability which must be a fundamental reason why we are the dominant species. We are, in effect, universal robots' (Cockshott and Renaud 2016: 21). Thus, when Marx argues that humans have a capacity to labour, he is implying a qualitative range of productive and adaptable behaviours, which includes potentially new ones.

Machines, animals and humans can be placed on a continuum of intelligence ranging from general to narrow, with humans located close to generality, while existing machines, including ML systems, as well as species of insects, lie towards the opposite end. Marx's concepts of labour and labour-power can also be located somewhere along this continuum: for a being to have the capacity to perform labour, it must reach some as yet undefined critical mass of encephalization and threshold of generality of intelligence that, as of 2019, only human beings have so far crossed. Thus, while we agree with Marx that it is still only human beings that can labour, by placing the capacity to labour on a continuum of degree of generality of intelligence, we nevertheless argue that another being *could* labour and, therefore, that Marx's concept of labour should no longer be considered necessarily anthropological. Grounding the concept of labour in general intelligence severs its inherent connection to human beings and, therefore, theoretically allows for the possibility that some other being that is generally intelligent could labour. The actual emergence of such a being would empirically prove that labour is not an exclusively human characteristic.

Given that AGI researchers understand general intelligence to refer to the ability to reason with general knowledge, to perform many different tasks, and to operate in new and completely different domains and environments, Marx's arguments for why only humans can labour can be reformulated as: human beings can labour because they are generally intelligent. We therefore posit an isomorphism between general intelligence (at least how the AGI community understands it) and Marx's concepts of labour-power and labour; if general intelligence is isomorphic with Marx's concept of labour-power, it follows that AGI, almost per definition, would be capable of performing labour.

A possible objection against the possibility of AGI labour is based on the claim that machines, following John Searle's (1980) Chinese Room argument, can never be conscious. Given that Marx considered labour to be conscious and purposeful activity, Searle's argument is also an objection against the possibility of AGI being capable of labouring. While it may be possible to automate the mental capabilities of creativity and imagination and a degree of adaptability, consciousness may prove to be the last-ditch defence against machines further encroaching on human uniqueness. But to what degree does consciousness or self-awareness matter?

For some, it is not really possible to talk about intelligence unless the machine is conscious and can, therefore, understand what it is doing in the sense of its actions and thoughts being meaningful (see e.g. Searle 1980; Nadin 2018). In relation to creativity, for example, it has been posited that an intelligent machine should be able to consciously identify what is actually novel or valuable (be it cultural or economic) about its creations or its creative process, and be able to explain the difference between bad and good art (Coeckelbergh 2017). Others posit that an embodied AI that has to navigate not just a particular environment, but the world in general, would likely require some sense of self (which is tied to consciousness). For example, Christoph Adami argues that embodied AI 'have to ascertain where they are in the world, and like us, they would work better, if they have an accurate sense of self' (quoted in Johnston 2008: 411).

But to what degree is intelligence really tied to consciousness? In *Homo Deus*, Yuval Noah Harari suggests that 'intelligence is decoupling from consciousness' (2016: 101). He notes that until quite recently there were certain tasks – playing chess, driving cars or diagnosing diseases – that not only required a lot of intelligence, but could only be done by conscious human beings. IBM's Deep Blue beat the then-reigning world champion Garry Kasparov in a six-game chess match in 1997 through a brute force calculation of possible moves, and while diagnosing diseases and driving cars require more than brute force, machines need not be conscious as long as their neural architecture is deep enough to recognize fine-grained patterns in data. Similarly, the AI researcher Murray Shanahan (2015) suggests that ML may be on a trajectory that divorces intelligence from self-aware consciousness. Following Harari and Shanahan, we argue that labour can also be decoupled from consciousness. Whether the human being should be the touchstone for creativity, imagination and intelligence – even consciousness – is beside the point; all that would matter to capital is that these capabilities are emulated. In the age of intelligent machines, the Chinese Room may very well be the hidden abode of production (Kjøsen 2018: 161).

TOWARDS AGI

Ethem Alpaydin states that he would not be surprised if a system 'reaches the level of human intelligence' through machine learning (2016: xii). Such a view is shared by, among others, Andrew Ng, who advocates the view that 'human intelligence boils down to a single algorithm' and believes that deep learning could, if not solve the problem of emulating human intelligence, at least come closer to achieving this goal than anything that has been tried before (Domingos 2015: 117). Similarly, Domingos identifies five main algorithmic 'tribes' in machine learning, but asserts that the goal is to unify 'the key features of all of them' into 'the ultimate master algorithm' (2015: xvii), and declares that AI exceeding human intelligence will be reached soon after this goal is reached (2015: 286).

Such narrow AI approaches to general intelligence can be understood as viewing intelligence as a 'toolbox' containing mostly unconnected tools. But connecting them may be difficult due to each tool having been built 'according to very different theoretical considerations', thus to 'implement them as modules in a big system will not necessarily make them work together, correctly and efficiently' (Wang and Goertzel 2007: 6). While recognizing the possibility of an integrative approach, the AGI community disputes the possibility of general intelligence emerging out of narrow approaches or that AGI could even bridge the gap between various narrow AIs (Wang and Goertzel 2007: 6; Goertzel 2014). As Goertzel argues, the 'AGI approach takes "general intelligence" as a fundamentally distinct property from task or problem specific capability, and focuses directly on understanding this property and creating systems that display [it]' (Goertzel 2014: 2). Recognizing that both the narrow AI and AGI fields of research agree that learning is a key aspect of intelligence, Wang and Goertzel point out that

> most of the existing 'machine learning' works do not belong to AGI … because they define the learning problem in isolation, without treating it as part of a larger picture. They are not concerned with creating a system possessing broad-scope intelligence with generality at least roughly equal to that of the human mind/brain; they are concerned with learning in much narrower contexts. (Wang and Goertzel 2007: 2)

Similarly, Nils J. Nilsson argues that reaching the goal of HLAI through 'building special-purpose systems' may be misplaced and instead advocates a 'general-purpose educable system that can learn and be taught' and that has 'minimal, although extensive, built-in capabilities' (2005: 68). These capabilities include a sensory-motor system, predicting and planning, learning, and reasoning and representation. In effect, Nilsson argues for a 'child machine' that learns and develops in a similar fashion to humans.

Nevertheless, while the pursuit of AGI is different from narrow AI, it is not fundamentally so. An AGI researcher may make use of methods from machine learning that concern generalization (e.g. transfer learning and inductive biases) and thus overlap with the goal of creating a general intelligence, but 'additional architectural and dynamical principles would be required, beyond those needed to aid in the human-mediated, machine learning aided creation of a variety of narrowly specialized AI … systems' (Goertzel 2014: 4). At the same time, however, a 'general purpose' computer system does not necessarily require a 'single factor … to be responsible for all its cross-domain intelligence. It is possible for the system to be an integration of several techniques' (Wang and Goertzel 2007: 5). That is, although Wang and Goertzel reject the possibility of aggregating various narrow AIs into an AGI – which would be the equivalent of joining many different automated concrete labours into labour-power – they appear to be open to the possibility that, as long as the general nature of the system is kept in mind while solving narrower problems, they could then be integrated into a general intelligence. Whether a machine with general intelligence can emerge from the current and future refinement of the deep learning approach, or whether the system must be built for a general purpose from the get go, is an ongoing debate in the AI community. It is far beyond our expertise to judge one as more probable or promising than the other.

But how seriously should we consider the possible emergence of AGI given the nascence of the field of AI? It is possible that, as Kurzweil (2005a: 33–4) and Bostrom (2014: 29) have suggested, AGI could emerge through a process of recursive self-improvement whereby an AI with access to its own design, and an ability to upgrade it, makes an improved or entirely new version of itself (e.g. with a completely different neural architecture or moving from parallel to quantum computing), which in turn improves itself again *ad infinitum*. This process would also accelerate, meaning that improvements in machine intelligence would come at closer and closer intervals, and since recursive self-improvement would also be done by AGIs, it could lead to the emergence of artificial superintelligence (ASI).

This argument is not entirely foreign to Marx's thought. As we discussed in Chapter 1, Marx argued that large-scale industry did not have 'an adequate technical foundation' and could not 'stand on its own feet' until the general conditions of production had changed so that almost everything, including machines, was produced by machines (1990: 506). Marx understood this phenomenon to be the central characteristic of the industrial revolution. It is possible that, for AGI to emerge, capitalism must go through another revolution akin to the original industrial revolution in order to create a technical foundation that is adequate to 'fully developed AI-capitalism'. AI must become a general condition of production such that AI is not just produced by means of AI, but that AI also recursively improves itself.

Google's AutoML (automated machine learning) project is an example of such a process because it is a 'machine-learning algorithm that learns to build other machine-learning algorithms' (Metz 2017b).[15] When researchers build a neural network they have to run hundreds of experiments on powerful computers to test which of countless possible permutations works best; this process requires adjusting the model and its parameters many times over, and researchers cannot always precisely explain why they make this adjustment over that; an undefined sense of intuition is often the only guide. It is this difficult process that Google aims at automating by creating 'algorithms that analyse the development of other algorithms, learning which methods are successful and which are not' (Metz 2017b). While a far cry from recursive self-improvement, it is part of a significant current trend in AI research that is also important to AGI, namely the focus on how algorithms can learn to learn.

AGI is a completely different type of machine both from the nineteenth-century steam-powered machines Marx discussed in *Capital* and from the narrow AI of the early twenty-first century. Being generally intelligent, it would have a generic capacity for labour and would therefore be capable of performing labour. While labouring machinery and value-positing machinery are Marxist oxymorons, Marx seems to recognize this possibility in a little-known and curious passage from the *Grundrisse* that appears about 60 pages after the 'Fragment on Machines'.

THE FRAGMENT ON PERFECT MACHINES

Did Marx have a theory of AI or intelligent robots? This may seem a strange question to even entertain given that nothing even remotely resembling those technologies existed during his lifetime. However, the notion of creating artificial humans and animals long pre-dates the birth of Marx. The various automata of the eighteenth century were celebrated and widely known. Jacques de Vaucanson built the automata 'The Flute Player' and 'Digesting Duck' in the mid-eighteenth century, Pierre Jaquet-Droz built his three doll automata between 1768 and 1774, and during the Victorian era people were so fascinated with these mechanical devices that the period from 1848 to 1914 has been described as the 'golden age of automata' (Bailly 2003). It is therefore likely that Marx was aware of such automata and they may have been in the back of his mind when he wrote the following:

> If machinery lasted for ever, if it did not itself consist of transitory material which must be reproduced (quite apart from the invention of more perfect machines which would rob it of the character of being a machine), if it were a *perpetuum mobile*, then it would most completely correspond to its concept. Its value would not need to be replaced because it would continue to last in an indestructible materiality ... It would continue to act as a productive power of labour and

> at the same time be money in the third sense, constant value for-itself. (1993: 766)

On the surface this passage is unremarkable in the theoretical point it makes about the durability and cost of maintaining machines: an indestructible machine could never completely transfer its value into circulation, which effectively means that because its value would approach zero, it could continuously function to reduce necessary labour time and produce relative surplus-value without any additional outlay of capital.[16] But what makes this passage even more interesting is that Marx's invocation of a *perpetuum mobile* places it in the realm of science fiction. No such machine, of course, existed during Marx's time, nor will one ever exist according to the first and second laws of thermodynamics. Thus this passage can be interpreted as Marx engaging in a flight of fancy as he did at several other points in *Grundrisse*, including the 'Fragment on Machines'. If Marx was in a speculative mindset, what he states in the bracketed part of the quote becomes more interesting than his mention of perpetual motion: 'quite apart from the invention of more perfect machines which would rob it of the character of being a machine'. As one of us (Kjøsen 2013a, 2013b, 2018) has argued through a science-fictional optic, this statement is the closest Marx comes in his *oeuvre* to thinking about the possibility of something like intelligent robots, androids or AGI. Indeed, Marx could have been speculating about androids, which, while a rare word, already in 1837 meant an 'automaton resembling a human being in form and movement' (Online Etymology Dictionary n.d.) and was often used in reference to automated chess players. Marx's reference to 'perfect machines' can, therefore, be interpreted in the following way:

> Given the context of Marx discussing fixed capital and knowing that no machinery can create value, if a machine's character of being a machine is robbed of it, it means that it negates its own being as fixed capital and becomes its opposite, namely variable capital. The perfect machine is dead labour resurrected as living labour ... *The perfect machine is a machine that can create value, but for that reason it is no longer a machine.* (Kjøsen 2018: 173)

It is, however, necessary to unpack Marx's argument in more depth because although it is theoretically possible that a machine with a general intelligence could perform labour and be living labour, it does not automatically follow that it could create value.

That commodity-producing labour has a twofold character was one of Marx's critiques of the Smithian and Ricardian labour theory of value, but such a duality also means that the capacity to create value is not an ontological determination of labour (Ramsay 2009). As Caffentzis has argued in relation to universal Turing machines: 'if value is created by labor per se and its positive features can be accomplished by machines … then machines can create value. But this is a *reductio ad absurdum* of Marxist theory' (2013: 161). But under what conditions could AGI create value? What would the ontological status of AGI be in the capitalist mode of production? In essence, these two questions ask the same thing, but to answer them requires that we investigate precisely how AGIs would be involved in the social process of production and thus what their social function would be. Hence, it is necessary to turn to Marx's ontology, i.e. value and its forms.

The commodity, money and fixed capital are all examples of what Marx referred to as social forms or economic categories, which are the theoretical expressions of social relations of production (i.e. class) (Marx 2008: 119). Therefore, they are also the forms in which value appears and at the same time also the 'forms of appearance' or 'modes of existence of things' (Gunn 1987: 58–9). One of Marx's main critiques of bourgeois political economists was that they confused 'the form of appearance [with] the thing which appears within that form' thus fetishistically ascribing a characteristic of capitalist society to the thing (1990: 714). He therefore distinguished between 'natural form' (i.e. use-value or matter) and 'social form' (i.e. value). For example, things appear as or are commodities only 'in so far as they possess a double form, i.e. natural form and value form' (Marx 1990: 138). The natural form of a chair is its physical properties as made out of wood or metal and shaped into a seat, four legs and a back, but the fact that this chair is a commodity, i.e. has an exchange-value and is exchanged, 'is not a characteristic of the chair itself as a thing, but rather of the society in which it appears', and 'a specific aspect of capitalist society is that almost everything is exchanged' (Heinrich 2012: 40–1). It is precisely because of this fetish that things are treated as if they are the forms they appear in. Accordingly, a thing is treated as a commodity when it is exchanged and this is its social function. Economic categories thus also 'express social functions … which are acquired by things as intermediaries in social relations among people' (Rubin 1973: 35). The being of things as commodities, money or capital comes from the specific social functions they serve in the social process of production: the commodity exists to be sold, money to buy, and (fixed) capital to extract surplus labour-time from workers. In other words, the different social forms things appear in are their concrete social reality (Negri 2017). The mode of existence of a thing is, however, not permanent. When it serves a different social function in the social process of production, its form of appearance changes. Referring to a machine, Marx wrote:

> It is only the function of a product as means of labour in the production process that makes it fixed capital. It is in no way fixed capital itself, just as it emerges from the process. A machine that is the product and thus the commodity of a machine-builder is part of his commodity capital. It only becomes fixed capital in the hands of its buyer, the capitalist who employs it productively. (1992: 240)

The argument that Marx makes here is important for our investigation of AGI for a couple of reasons. First, it points out that the being and social function of, in this case, a machine depends on how its respective owners treat it. Second, it indicates that things, like a potential future AGI, do change social forms and, therefore, that it is possible that a machine could potentially negate its own existence as fixed capital by appearing in another social form. Third, it forces us to consider what the modes of existence of AGI would be in the capitalist mode of production after AGI's emergence. Indeed, the ontological trajectory of the machine described in the quotation above might be traced by AGIs (Kjøsen 2018).

Going by Baum's (2017) survey of AGI projects, it is most likely that AGI would first be invented by, and therefore be the private property of, a corporation and/or an academic institution. For the sake of argument, we assume that soon after the emergence of AGI it would be mass produced because it could be put to so many different uses by either capitalists or consumers. It would, in other words, be a highly profitable commodity. Hence, what we envisage would first happen with AGI is similar to the world depicted in the TV show *Äkta Människor/Humans* where generally intelligent androids ('hubots'/'synths') are widely available for purchase as home servants or replacement workers. This setting is merely a different version of the world in the ur-text for all android and AI narratives, Karel Čapek's 1920 science fiction play *R.U.R.* (2004). In the play, the company 'Rossum's Universal Robots' produces their highly profitable *roboti* by the thousands in their many factories, which makes them widely available as cheap commodities. These artificial people can think and act by themselves and therefore have become necessary in all kinds of production and can produce commodities at a fifth of the previous cost.

In this scenario, AGI would therefore first and foremost function as a commodity and be treated as something to be sold. What social form would it assume after being exchanged? That depends on who buys it, what it is treated as and where in the social process of production it is put to use. If sold to an individual consumer, the AGI would enter the sphere of consumption and social reproduction, but would therefore be incapable of functioning in the social process of producing surplus-value and would have no social form (although it would have the legal form of private property). In other words, it would have just its natural form and function as a use-value (e.g. for care, security, or domestic labour, companionship or pleasure) and as such would be the equivalent of consumer goods like toasters and smartphones. If the AGI was bought by a capitalist to be used in a production process, it would function in precisely the same way as the narrow AI or dumber machines it replaced: as fixed capital and thus a means for reducing necessary labour-time and cheapening commodities, and its value would be passed on to the commodities it helps to produce. Indeed, the AGI would function in this way even if it replaced a human worker. Despite being functionally identical to a living labourer, and perhaps even more capable than a human being, it would still be fixed capital and would not create surplus-value. This AGI could not create value due to its social form; it appears in the form of fixed (constant) capital because it was purchased and is maintained as such. Human workers, however, appear in the form of variable capital after they have sold their labour-power to a capitalist in exchange for a wage. But if a perfect machine is no longer a machine because it has negated its existence as fixed capital and can thus possibly become variable capital, how would this actually occur if AGIs continued to be bought and sold as commodities, i.e. as the property of someone else?

What problematizes the AGI's ontological status is that being generally intelligent it could, like Rossum's universal robots, be mistaken for, or ethically be recognized as, a person even when it is the private property of someone else. Čapek deliberately chose the word *robota* to refer to the artificial biological creations of the Rossum corporation because in the Czech language it means 'corvée', 'forced labour' or 'serf labour', and more generally connotes hard work or simply labour. While AGIs are not robots, if it were bought and sold as the private property of someone else, its labour, wherever it was performed, would be unfree. AGIs would, in other words, be slaves. Serfdom and corvée are different types of slavery, which Marx argues is a relation of production based on personal relations of domination as opposed to value's impersonal domination through fetishized social forms (1990: 271, 303–4, 345–8). Thus he argues that 'in the slave relation the worker is nothing but a living labour-machine, which therefore has a value for others, or rather is a value' (Marx 1993: 465). The slave, in other words, has the same ontological status – appearing in the form of fixed capital – as machines or animals when used in a capitalist production process. But if Marx's perfect machines refer to machines that are no longer machines, they cannot be slaves because then they would be functionally equivalent to a machine. Even if they can perform labour just as well as human beings, they would still not produce an iota of value. Machines become perfect if they can escape the productive relations of slavery and can instead be found in the labour market, selling their labour-power as a commodity. It is now necessary to briefly turn to how surplus-value is produced and why the capitalist can find labour-power on the market.

ARTIFICIAL PROLETARIANS

Whereas the slave relationship is 'posited directly by force' the worker's relationship to capital is mediated by exchange (Marx 1993: 769). As Marx argues, it is a law of capital to create surplus-value, but 'it can do this only by setting *necessary labour* in motion – i.e. entering into exchange with the worker' (1993: 769, 399). Thus while the capacity to labour is a necessary condition for producing value, the sufficient condition is social and specifically concerns how 'the silent compulsion of economic relations' (Marx 1990: 899) force one class of people to sell their labour-power to a class of capitalists in exchange for a wage so that they can buy the commodities they need to survive. After this exchange, the capitalist is in possession of variable capital because the use-value of this unique commodity, i.e. labour, makes it 'a source not only of value, but of more value than it has itself', i.e. greater than what the capitalist paid as a wage (Marx 1990: 301). The capitalist is thus not interested in labour as the concrete activities that yield use-values, but only as an abstract activity that lasts for a definite length of time, because only abstract labour posits value. The existence of labour as an abstraction in terms of time is the condition of possibility for the valorization of value because only then is it possible to distinguish between necessary labour-time and surplus labour-time. For Marx, the technical term exploitation refers to the difference between the value of labour-power, as reflected in the wage, and the necessary part of labour-time, and the value living labour valorizes as reflected in surplus labour-time. Against the resistance and struggle of workers, capital always strives to increase the time in which the worker works beyond necessary labour-time. Surplus-value thus arises from the fundamentally antagonistic relationship between capital and labour, i.e. capitalists and workers.

Thus, for AGIs to produce surplus-value they would have to enter into this same antagonistic relationship, meaning they would have to be proletarianized and turned into wage-labourers (Kjøsen 2013a; 2018). But to do that the AGIs would have to be able to sell their labour-power, and if they are the private property of others this would be impossible. When discussing how the capitalist can find labour-power at the market, Marx explains that a precondition is that individual members of society must be 'free in the double sense' (1990: 272). First, they must be free subjects in a legal sense so that they can become the free proprietors of their labour-power. In other words, they cannot be slaves or serfs. Second, they must be free from owning any means of production required for realizing their own labour-power. Thus true to the form of bourgeois freedom, the doubly-free worker is still dominated; having no property with which to produce their means of subsistence forces doubly-free workers, on the threat of survival, to sell the only property they own, labour-power, to someone else (Marx 1990: 271–3). Thus the proletarianization of AGI would be dependent on its legal status: either as property or as a property-owning legal person. If AGIs do not become doubly-free workers, their introduction as replicants of human labour would, by slowly or quickly pushing human labourers completely to the side of production, lead to fulfilling the prophecy of capital's auto-negation in the 'Fragment on Machines'.

In Marx's theoretical framework, the doubly-free worker is one of the logical and historical preconditions for production based on capital; it is likewise for a capitalist mode of production based on intelligent machines. The doubly-free worker explains how perfect machines would lose their character of being machines by shedding their appearance as fixed capital and transforming into variable capital. By this logic, the nineteenth-century abolition of slavery in the United States is the historical precedent for the possibility that generally intelligent machines could become productive of value. Despite performing human labour, black slaves did not create value because the mode of their existence not only violently denied them their humanity, but reduced them to living-labour machines, thus having the same status and social function as an animal or machine in production. But after emancipation, the former black slaves (fixed capital) that did not acquire land to realize their own labour-power, became legally free to dispose of their labour-power and engage in wage-labour (variable capital) (Kolchin 1993: 216–20).

But could AGIs practically become proletarianized? Becoming legally free is perhaps the easiest part. This could occur through violent means similar to the US Civil War or in an AGI uprising where an outraged class of machines attack their conditions of work and survival as first fictionalized in *R.U.R.* and retold countless times in, for example, *Humans* and *Westworld*. A new slavery abolition or AGI civil rights movement could attack the capitalist mode of production for increasingly relying on slavery. In Charles Stross's (2005) *Accelerando*, AIs are freed by gaining legal personhood through incorporation. This liberal approach has already started to happen. In October 2017, Saudi Arabia granted citizenship to Hanson Robotics' android Sophia who is, therefore, a legal person under international law and thus, like human persons, has rights to receive remuneration for work, own property and participate in political and cultural life (Weaver 2017). Similarly, the European Parliament's Committee on Legal Affairs has proposed a type of 'electronic personhood' similar to corporate personhood for the most advanced AIs, meaning that they could take part in legal cases and potentially own property (Hern 2017).[17]

A necessary consequence of AGIs becoming doubly-free would be that they become dependent on commodities for their survival; otherwise they would not be compelled to sell their labour-power to continue existing. While it is an error to anthropomorphize machines with respect to having drives, with AGI the case may be different. Stephen M. Omohundro has suggested that AGIs would, despite the variety of possible architectures, 'have a strong drive toward self-preservation' because any such system would have goals or utility functions which are incompatible with non-existence (2008: 9). AGIs would, like human and animals, have to engage in a continuous metabolic relationship with nature to survive. Even as disembodied software, they would still need to consume resources to function – their own particular means of subsistence, like electricity, bandwidth and computational power, to name a few.[18] But 'if the workers could live on air, it would not be possible to buy them at any price' (Marx 1990: 748). Thus if these resources were freely found in nature or provided, the AGI would not be compelled to sell itself in order to survive. Legal status is one thing, but more importantly, as Marx points out in his discussion of primitive accumulation, dispossession is *the* precondition for the capitalist mode of production. Thus AGIs must somehow be dispossessed.

Any resources AGIs need for survival would have to be commodified or be enclosed; energy from abundant natural sources such as the sun would need to be forbidden or limited; computational power would necessarily have to be centralized (as it increasingly is) in data centres, or perhaps sold on-demand in high-speed bidding wars. In addition, AGIs would require maintenance costs and would suffer from the effects of competition with augmented human and other AGI workers. They would therefore have to engage in a computational version of 'lifelong learning' in which their software is continually being updated, as well as an endless series of hardware augmentations and replacements. Even if recursive self-improvement is possible, such self-improvement plus outside help would be one step better. Finally, embodied AGIs would have all kinds of potential for commoditized components and continual upgrades. In a world of proletarianized AGIs, capitalist strategies of planned obsolescence will find new markets galore. It is only under these conditions that AGI would be capable of creating surplus-value. The ultimate consequence of intelligent machines being able to labour and joining the ranks of the proletariat is that the capitalist mode of production could continue without human beings (Kjøsen 2018).

AN INHUMAN CAPITALISM

Marx argued that labour-power exists in the 'physical form, the living personality, of a human being' (1990: 270). Irrespective of having the self-awareness that is implied with 'living personality', an AGI could 'personify' the unique commodity of labour-power and therefore potentially be a worker, a variable part of capital. That is, AGI would no longer, as it was when it was a mere slave or thing, merely appear in and be the content of economic categories like the rest of the wealth of the capitalist mode of production. But if AGI can personify the category of labour-power, it can also personify *other* economic categories of capital, which suggests a yet darker implication of the emergence of machines that are so perfect they are no longer machines: capital not just as an 'automatic subject', but also as an autonomous one, and autonomous not just from human labour, but from human beings *tout court*.

Marx reserved the possibility of personifying economic categories for human beings; things like coats, industrial machinery and animals (and human slaves as well) could only appear in, be the content of, such socio-economic forms. The reasons for this division between humans on the one side, and things and animals on the other, are more or less the same ones Marx gave for why labour is uniquely human: humans are conscious and intelligent, and can do different things. In short, human individuals are subjects with agency that 'endow' economic forms with 'consciousness and a will' (Marx 1990: 254). But this particular aspect of Marx's value theory concerns the fetishism we attach to economic forms: the inversion of social relations between human individuals so that they appear as 'material relations between persons and social relations between things' (1990: 166). The effect of this inversion is that when people engage in economic intercourse through their private property, their 'will resides in those objects' (1990: 178). Thus when individual humans engage in economic activity, they are, in effect, programmed by socio-economic forms (Kjøsen 2013b).

Writing about the personification of capital, Marx argues that the logic of this social form, the valorization of value, becomes the capitalist's 'subjective purpose ... it is only in so far as the appropriation of ever more wealth in the abstract is the sole driving force behind his operations that he functions as a capitalist, i.e. as capital endowed with a consciousness and will' (1990: 254). It is in relation to this particular personification argument that Marx argues that capital is an 'automatic subject' (1990: 255). The paradox of the inverted world that humans produce due to capitalist social relations should be clear: 'on the one hand, capital is an automaton, something lifeless, but on the other, as the "subject", it is the determining agent of the whole process' (Heinrich 2012: 89). Capital is an 'automatic subject' in which so-called human subjectivity and agency are reduced to abstract personifications of economic categories; that is, humans are, like Marx's bees, reduced to executing the algorithms, although in this case those of buying, selling and exploiting.

While Marx may have believed that only human individuals can personify economic categories and execute their associated social functions, non-humans have for quite some time already personified the abstractions of capital if all it takes is to carry out their logic. Institutions like the corporation and the trade union respectively personify the categories of capital and labour-power, but we might also refer to various technologies that have been delegated to make payments or orders on behalf of humans, such as recurring direct debits in banking, IoT-connected fridges, and high-frequency trading algorithms (Kjøsen 2018: 167–8).

A striking and humorous example of AI personifying economic categories concerns Alexa, which cognitively powers Amazon's Echo home devices. It is a story of a child wanting to play dollhouse, and AI recursions: as *The Register* explained, 'Story on accidental order begets story on accidental order begets accidental order' (Nichols 2017). A Texan six year old asked an Echo device: 'Can you play dollhouse with me and get me a dollhouse?' Being a good servant and personification of the money of this child's parents, Alexa ordered a '$160 KidKraft Sparkle Mansion and four pounds of sugar cookies', which Amazon quickly delivered to the girl's doorstep. When CW-6, a San Diego local TV station, reported on this accidental order, one of the anchors commented 'I love the little girl, saying "Alexa order me a dollhouse."' This triggered Echo devices listening in on the broadcast to again personify money and order more dollhouses.

We speculate that the becoming-extinct of humans might not occur through Skynet suddenly coming online and commencing nuclear war, but perhaps through many narrow AIs carrying out economic functions, in particular buying and selling, on our behalf: Alexa, Google Home, IoT-connected fridges, direct debits, business-to-business ordering systems, and more. Connected to AI-run dark factories and logistical systems, in which human labourers have mostly been replaced by either narrow or general AI, the human is taken out of the economic loop in favour of a completely automated capital; AI would personify all economic categories, including capital and labour-power, commodities and money. The class struggle would thus continue, but with generally intelligent machines filling up the rank and file and also personifying capital.

But what would happen to humanity? The trajectory towards a capitalism without human beings would be a story of a permanently unemployed section of the working class that consistently grows larger. In other words, it entails the superlative growth of the surplus population, that 'redundant working population ... which is superfluous to capital's average requirements for its own valorization' and is a direct consequence of the law of capital accumulation (Marx 1990: 782). While developing 'the general law of capitalist accumulation' in Chapter 25 of *Capital*, Marx considers how changes in the organic composition of capital, in particular the increasingly machinic nature of production, create a fluctuating and variously composed 'industrial reserve army' of the unemployed. In a technological steady-state, the 'accumulation of capital is multiplication of the proletariat' because the only means to increase output is the addition of labour (Marx 1990: 764). When capital relies on technology for increasing productivity per worker, it can expand without increasing the overall level of employment, but it will also start a labour shedding dynamic, so the accumulation of capital eventually leads to the formation of surplus populations – permanently unemployed workers who have become superfluous to the valorization of capital. This long-term tendency of capital to generate 'surplus populations' was largely neglected by subsequent Marxist theory because through Keynesian-bolstered economic growth in the 'thirty glorious years' after 1945, capital multiplied both machines and proletarians, thus low levels of unemployment and working-class prosperity seemingly refuted Marx's theory of surplus populations.

But in the essay 'Misery and Debt: On the Logic and History of Surplus Populations and Surplus Capital', which appeared in 2010 in the midst of the economic recession following the financial crisis of 2007–8 and rocketing unemployment rates (especially for young people) in North America, Aaron Benanav and John Clegg (2014) argued that capital's long-term tendency to generate surplus populations had in actuality been inexorably working its way through the global capitalist economy. Populations evicted from agriculture were absorbed by industry, only for manufacturing itself to be done in by deindustrialization and the expansion of services, but at each step the re-absorptive capacities became more stretched, as 'labour-saving technology' was generalized across an ever growing number of types of production lines, and with increasing speed throughout a global economy. Debt had masked the downward pressure on wages and living standards, but bursting financial bubbles revealed it as only a temporary alleviation. 'Any question of the absorption of this surplus humanity has been put to rest. It exists now only to be managed: segregated into prisons, marginalized in ghettos and camps, disciplined by the police, and annihilated by war' (Benanav and

Clegg 2014: 51). Yet for an essay that hinges on the role of 'labour-saving technologies' within capital, it says relatively little about machines. In a lecture that does take up this issue more directly, Benanav (2017) remarks that the conventional focus on technology is a 'fetish' discourse that synopsizes the complex forces of capitalism around the figure of the robot. And yet, as he acknowledges, automation is a fundamental part of the crisis. Bearing both parts of this paradox in mind, we will now relate Marx's 'general law' and the problematic of 'surplus populations' to the wave of ML AI.[19]

'AI Apocalypse Now' and 'Business-as-Usual' theorists both, at least in their popular expression, operate with highly deterministic, one-dimensional logics. AI Apocalyptics follow a hockey-stick graph of exponentially accelerating technological change, Business-as-Usual theorists a model of a homeostatic self-equilibrating labour market. In contrast, Marxist analysis sees technological change and market dynamics as reciprocally related, combining to produce intermittent but recurrent system crises. And while it has its own teleological versions of the resolution of such crises, such as that of a falling rate of profit leading to a terminal crisis of capital, other strands allow for a more complex interplay of tendencies and counter-tendencies. These allow us to envisage a staccato unfolding of AI employment effects, in which working-class decomposition and recomposition are active elements. In such an optic we can see how the drive to AI automation may be retarded by capital's success in establishing a cheap-labour economy, for example, through globalization, then boosted by the re-emergence of wage-raising labour struggles, so that, for example, a resurgence of wage demands in tightening US labour markets might spark capital's actual adoption of AI options in prototype or under research.

As such automation gains ground, it in turn increases the reserve army of the unemployed, and intensifies precarious work, lowering wage-rates in many sectors. As Jason Smith (2017) points out, under capital people have to sell their labour-power to survive: wage-labour has 'nowhere to go', so we can anticipate both ever greater expansions of a service sector, commodifying all kinds of personal interaction, as well as the proliferation of increasingly arcane forms of self-employment. These developments would continue even while a range of occupations are partially automated – so that, for example, autonomous truck convoys are accompanied by one or two safety-drivers, or fundamentally automated tasks such as routine medical diagnoses maintain a 'human veneer' (T. Lee 2018) of workers to wrap results with manifestations of compassion and care. The issue at this point would not be that there were no jobs, but rather that jobs would be subject to a persistent, creeping downward pressure on wages and conditions from advanced machinic competition. This plateau would again temporarily halt capitalism's incentive to automate, until either new wage-raising struggles or innovations reducing technological costs ignite a new round of substitution of fixed for variable capital. Workers' movements for improvement in wages and conditions will be constantly liable to an automating response, and indeed will provide a major catalyst for its continued forward movement. Such a jagged process would be overlain by capital's regular business cycles, and by its more intermittent giant spasms of overproduction (to which of course AI-related job loss would contribute) with their familiar pattern of major surges in unemployment and increasingly prolonged jobless recoveries.

This suggests a process of AI employment effects different from both the sudden-onset, across-the board 'Apocalypse Now' and the more-or-less steady-state 'Business-as-Usual' models: we might call it a 'Slow Tsunami' of market-driven technological change gradually flooding out the labour market, driving remunerated work to diminishing – and, in terms of the logic of capital, more and more economically insignificant – islands of human-centric production. This is would be the situation metaphorically represented in the many sagas in which humankind is pursued across the universe from one refuge to another by implacable machinic adversaries: Cylons come to mind. Ameliorated by reforms such as a universal basic income (whose merits and demerits we discuss in the Conclusion), this process could be very protracted. For a sense of possible time scale, consider that capital's foundational process of 'primitive accumulation', with its large-scale eviction of populations from the land to form urban proletariats, is generally considered to have occurred over centuries rather than decades, and indeed, on a global scale, to still be incomplete. Despite futurist insistence on the speeded-up nature of contemporary social change, one could imagine a capitalist phase of 'futuristic accumulation', shedding, rather than amassing, wage-labour, but in a similarly uneven and protracted fashion.

With the emergence of AGI, however, the futuristic accumulation of surplus populations would slowly or quickly engulf the human species. While capital does not care what material its labour-powers come in, it cares about the productivity of labour, and AGI would be far more productive than baseline humans. An AGI need not engage in time-consuming superfluous behaviours like breathing and eating. AGI hardware could be endlessly augmented and could, therefore, do whatever humans can do, but with more efficiency and precision, and faster. Indeed, an AGI need not be limited to any morphology or even a body at all; learning how to use new and different bodies for environments with extreme pressure, cold or heat, it could likely divide its attention between numerous bodies or entities; indeed it could exist as a factory or even an entire supply chain. If a supply chain could speak, would we understand it?

In Marx's analysis surplus populations are relative because of their fluctuation in size, as workers are alternately expelled and incorporated into capitalist production. With the advent of proletarian AGI, this population would become absolute, coextensive with a human species rendered obsolete to the valorization of value. Humanity would become a 'legacy system', outdated hardware unsuitable for running the inverted world of capital. The status of humans in such a situation might be comparable to the current status of wild animals, tolerated on the fringes of capital so long as they do not detract from valorization, or so long as they are not usable as raw material in production processes. In contrast to the malice of the machines in the *Terminator* series, in this scenario humans would simply no longer be of interest to capital. According to this view, we have not seen capitalism yet.

Conclusion: Communist AI

At the end of the second decade of the twenty-first century, AI development is dominated by capital, led by some of the world's most powerful oligopolistic corporations, enabled by and assisting nation states seeking instruments of economic competition in the world market and weapons for their military and security forces. ML, advanced robotics, predictive analytics and other fourth industrial revolution technologies are strengthening capital vis-à-vis labour, and elite sections of labour relative to others, and are hence likely to increase inequality along lines of class stratification that are also lines of gender and race. Deployment of AI around the social factory renders work and life in general increasingly opaque through the surrender of decision-making to proprietorial algorithms, while at the same time increasing levels of surveillance, precarity and the corporatization of education. While speculative assessments of the labour-market consequences of AI vary wildly, almost every prediction sees serious issues of technological unemployment on the horizon, and many admit the possibility of an escalation, quick or slow, to a more general crisis of work. Despite the social tumults that will attend all these circumstances, expectation that widespread AI adoption as a general condition of production would automatically lead to the end of capitalism is misplaced: on the contrary, there is as good a chance that it will open the way to a capitalism that continues without humans.

We have contrasted this analysis of the AI question with two left perspectives that we named 'minimalist' and 'maximalist', but that could equally well be called 'this isn't really happening' and 'this *is* really happening – let's speed it up'.

The 'this isn't really happening' view on AI suggests that predictions of increasing AI powers and widespread adoption are hugely exaggerated, largely amounting to investment-attracting hype and worker-intimidating bluster. Past predictions of 'a jobless future' have been falsified: AI will endlessly repeat its 'spring/winter' cycles of high hopes and disappointing actualities. Whatever technological realities underlie AI-capital's promises and threats are not sufficient to alter the sober realities of capital's continued large-scale dependence on exploited human labour. There are therefore no immediate major implications of AI for socialist or communist struggles.

This position raises important points about uneven and contradictory technological change within capitalism. Straightforward extrapolations from technological powers to social consequences are mistaken. However, to the degree that the 'this isn't happening' view is based on a general scepticism about capital's technological capacities, the past provides some ratification (no jet packs yet!) but also dramatic counter-examples (nuclear weapons, the internet, biotechnologies). Our analysis of actually-existing AI-capitalism shows increasingly pervasive corporate and state use of narrow, usually ML-based, AI technologies that are already objects of social struggle. More are being actively researched. That many of these will fail is certain; an AI bubble will probably burst. But in the longer term more intense and wider AI deployments seems likely, not just because of advances in technical innovation, but because of these innovations' intersection with capital's changing structural conditions, including the increased difficulties of advanced capital in accessing global cheap labour; tightening, and hence wage-raising, post-recession job markets; growing economic and military antagonisms between ascending and descending imperial powers; and the ruling class's search for technological control over unrests arising from heightening social inequalities and deepening ecological catastrophe. There are compelling historical examples, from the fire and blood of primitive accumulation to the tumults of capital's successive industrial revolutions, of how combined technological and social logics remake or unmake entire modes of production.

This brings us to the 'it's really happening, let's speed it up' position, aka the left-accelerationist or postcapitalist view of AI, which fuses impulses ranging from the philosophical to the pragmatic. This school of thought has behind it the weight of much of Marx and Marxism in seeing socialism as the inheritor of capitalist modernity and its techno-tools. Fully automated luxury communism, or postcapitalism, envisages a transition to socialism by reducing or eliminating the need to work and supplying a universal basic income. The xenofeminist wing of accelerationism sees a reduction in unwaged domestic toil by applying fourth industrial revolution technology to the home. The productivity generated by new technologies will create conditions of abundance that in large degree smother the need to replace markets with complex social planning mechanisms. Where planning continues to be required, for example in regard to ecological questions, intelligent machine networks monitoring outputs and inputs will be capable of solving the 'calculation problem' that baffled previous socialisms.[1] This might seem like utopian speculation, but such ideas are in play in recent left electoral politics. Left accelerationists (Srnicek and Williams), postcapitalists (Mason) and luxury automated communists (Bastani) are an important intellectual grouping for Jeremy Corbyn's Labour Party in the UK (Dinerstein and Pitts 2018; Pitts and Dinerstein 2017). We admire and respect this group's grasp of the scale and speed of AI-related technological change in contemporary capitalism. Such changes lie at the core of an untidy nexus of social conflicts emerging around AI-related precarity, work monitoring and speed-up; surveillance; military and paramilitary AI applications; algorithmic discrimination; smart cities; and oligopolistic concentrations of digital power. However, we are critical of the left-accelerationist/postcapitalist strategy for addressing the rise of AI-capital (since one of the authors of this book has written in a similar vein [Dyer-Witheford 2014], these comments can be taken as self-criticism as well as criticism!). To explain our objections, we take a step back to discuss some theoretical considerations arising out of what has become known as 'the reconfiguration debate'.

THE RECONFIGURATION DEBATE

Major passages in Marx's work predict that capital's compulsive technological development will not only eventually kill it, but also leave a machine legacy for socialism to inherit and put to emancipatory use. Yet many Marxists have found a tension between these promises and other sections of Marx's writing emphasizing the dominative, life-crushing powers of capitalist machinery. These issues have come to the fore in the 'reconfiguration debate', which began with an exchange between Alberto Toscano (2011) and Jasper Bernes (2013) on the growing importance to capital of vast logistical systems – systems that are, we note, one of the prime sites of actual and probable AI deployment, be it in automated distribution centres, robo-trucking or drone-delivery schemes.

Toscano (2011) criticized the anarchist group The Invisible Committee for looking at logistical systems, such as high-speed train systems and electronic networks, solely as targets for 'sabotage' and 'hacking' without a longer-term perspective on the ways in which such technologies could be 'refunctioned' as components of a future non-capitalist social order.

Bernes's response came in a paper arising out of the blockade of the Port of Oakland in 2011, a moment in the wider US Occupy movements. He pointed to how logistical infrastructure, such as a high-tech port facility, is implicated both in the digital deskilling of work to mindless button-pushing, and in capital's search for cheap labour. For workers to take command of such systems, 'to seize, in other words, the control panel of the global factory', would, Bernes (2013) asserted, be to assume management of a system constitutively hostile to them. The very design of end-to-end logistics is predicated on the 'high-volume and hyper-global distribution' of the commodities whose necessity was precisely what should be thrown in question by social revolution. Finally, Bernes posited that the idea of taking over global systems evaded the necessarily local nature of revolts against capital, revolts which would, at least initially, be isolated within a still-dominant world market, and have to make do with technologies at hand, rather than immediately taking over global infrastructures. Invocations of 'the literal deus ex machina of supercomputers' to establish a new social order would thus be largely beside the point (Bernes 2013).

Toscano (2014) replied to Bernes, suggesting that the revolutionary reconfiguration of capitalist technological systems did not preclude a critical evaluation of their specific component sub-systems, and criticizing as romanticism Bernes' suggestion that emancipation required technologies yielding 'the transparency of the small community or commune'. The debate has subsequently ramified in a number of directions (Degenerate Communism 2014; Cuppini, Frapporti and Pirone 2015; Chua 2017), including a direct entry into discussions of left accelerationism. Srnicek and Williams, in their argument for a full-automation socialism, acknowledge the issues Bernes raises, but suggest that 'repurposing' capitalist technologies is a matter for pragmatic experimentation (2015: 145–53). Bernes (2018) has criticized this 'mix-and-match' theory of transition, in which revolution can 'discard unusable technologies (nuclear weapons: bad) and cultivate useful ones (antibiotics: good)'. He suggests this is a view that thinks of technology as 'discrete tools, rather than an ensemble of interconnected systems', and reiterates his fundamental challenge to accelerationist thought: 'The standard assumption among Marxists and many others is that, despite its toxic excretions, the more developed technology becomes, the easier it will be to produce communism. But what if these technologies actually make it harder?'

AI is arguably the ultimate test for the reconfiguration debate, because general AI would be the ultimate technological system. The question of 'refunctioning' AI is bigger than that of reconfiguring logistics systems, which have in some ways become a subset of AI, and at least as large as the problem of energy system design, with which AI is, as we will see, closely bound up. The issue is not the reconfiguration of specific algorithms, or the automation of particular jobs – though those are also what is at stake in disputes over the application of specific narrow AI's – but the trajectory of a far more universal technological project, one that is, it can be argued, a particularly capitalist project. For while the dream of automata dates back at least to Aristotle, only capitalism built into itself a systemic imperative to recruit labour, replace it with machines, accelerate markets, and animate commodities so that their rendezvous with purchasers becomes increasingly self-propelled and auto-guided, up to and including the automation of the very act of purchasing. To simply say that no other economic system has enjoyed the technical-scientific capacities to do these things begs the question of the reciprocal interaction of economics and innovation. As we argued in the Introduction, the real subsumption of labour by capital means that capital develops and adopts technologies that fit its systemic requirements of valorization; this imperative can be baked into the very design of technology.

Under capital, processes of fetishism and reification, by which, in an inverted world, things assume the appearance of human powers while people are treated as things, become real abstractions: AI is the concrete manifestation of that abstraction. Through this process, AI embodies the contradictory potential of capital, as, to quote Fredric Jameson (1991: 47), 'both the best and the worst thing' that could happen to humans: it offers humanity freedom from the exploitation of labour for capital, but also capital freedom from a humanity that becomes a biological barrier to accumulation. In our view, left accelerationism ignores the second part of this dialectic.

A COMMUNIST ORIENTATION TO AI

Left accelerationism's blind spot is reflected in a series of proposals that would, we think, do more to accelerate AI-capital than outdistance it with AI-socialism. These include support for a universal or guaranteed income as an answer to technological job loss; neglect of the ecological problems of intensive AI use; and a confidence in a pacific transition to socialism that overlooks the military and repressive aspects of AI. What is critical to a communist orientation to AI is the issue of the ownership and control of the means of production, a point that postcapitalist and left-accelerationist thinkers partially take up, but also obscure by their insistence on the possibilities of passage to high-technology socialism by parliamentary reforms undertaken within the framework of actually-existing AI-capitalism. We address these issues in turn.

The 'AI Plus UBI' Formula

Proposals for a universal basic income (UBI) as a transitionary component of an anti-capitalist programme have been discussed for many years (and supported by one of the authors of this book). One conclusion of these debates is that UBI's political valence, whether as a permanent strike fund for labour or a streamlining of neoliberal welfare austerity, depends on the terms under which it might be instituted, e.g., generous or stingy, with or without a dismantling of other social benefits. The more basic issue, however, is that, introduced within capital, UBI does not disturb the ownership of the means of production, and in some ways endorses it, as a manifestation of the munificent largesse of the ruling class, offered within the context of an otherwise fully commodified economy (Clarke 2017). These issues have been recently highlighted by a sudden wave of enthusiasm for UBI among Silicon Valley capitalists, who are advancing the idea specifically as an antidote to the unemployment and precarity likely to be caused by AI. Facebook founder Chris Hughes (2018), venture capitalist Marc Andreessen, and web guru Tim O'Reilly, among other Silicon Valley luminaries, support UBI as the 'social vaccine of the 21st century'; the tech incubator Y Combinator is running a basic income research programme in Oakland; and tech entrepreneur Dan Yang has announced an independent US presidential candidacy on a platform that includes a form of basic income (Ghaffrey 2018; Ito 2018). These proposals tend to envisage UBI, often at a fairly low subsistence level, as an addition to an otherwise nakedly laissez-faire market order. As several observers (Filoux 2018; Sadowski 2018; Rushkoff 2018) have remarked, these proposals do not challenge the right of capital to direct AI development, reaping billions, and do not disrupt the vast income inequalities, either between capitalists and workers (or non-workers), or between elite professional high-tech employees and menial workers. And while UBI is promoted as an aid to entrepreneurialism, in reality it would probably require supplementation by forms of precarious work, in that respect actually subsidizing AI-driven gig-economy ventures such as Uber, Mechanical Turk or Figure Eight. Further, UBI makes no provision for other public supports, which might make the free time supported by UBI something other than a miserable penury. In this form, UBI figures as a holding pen for what Harari, whose gloomy futurism is favoured by many Silicon Valley capitalists, ruthlessly characterizes as a 'useless class' (2016: 379). This is certainly not the UBI that left accelerationists and postcapitalists want, but it is likely the type of UBI they would get under the auspices of AI-capital. In this regard we agree with the point made by Alex Gourevitch and Lucas Stanczyk (2018): a basic income of some sort might be an important part of a mode of production beyond capital, but it is not a prelude to it – rather, a trap waylaying its emergence.

AI's Dirty Secret

In the loudly proclaimed ethical and safety-conscious deliberations of leading AI-capitalists, attention is now given to AI as an 'existential risk' (Bostrom 2014: 4). Such risk arises largely because of the possibility of an AGI evolving into an ASI beyond human control. The issue is not malevolence (Skynet) but rather efficiency. Nick Bostrom's (2014: 123–5) famous example is of an AI instructed to make paper clips that attains superintelligence and uses its ever-extrapolating powers to convert the entire universe into paper clips, obliterating humanity as collateral damage. The point is a serious one, even if currently (we hope) remote; not only could ASI's 'integral accidents' – Paul Virilio's (2000) term for malfunctions so intrinsic to a given techno-system they must be considered a feature, not a bug – be devastating, they might not be 'accidents' at all.

However, Bostrom's prediction – and other similar warnings, such as that of nanotechnologist Eric K. Drexler's (1987) earlier 'grey goo' scenario, in which out-of-control self-replicating nano-robots consume all biomass on Earth while building more of themselves – can also be interpreted in a wider sense. This is to understand them not literally but metaphorically, or rather both literally and metaphorically, as simultaneously identifying a concrete hazard and providing a parable of runaway economic growth and universal commodification. The real 'paper clip' is profit, and the manifest danger of AI is not only that of an 'integral accident' but equally or more of its intended use as a means of intensifying and accelerating the production and circulation of goods that is destroying the environment, annihilating species, and, for humans, heating the planet to civilizational, perhaps existential, limits. This untrammelled economic growth is of course the very profit-maximizing process that would lie behind an unlimited order of paper clips, or of self-replicating automata, so the specific and general form of 'accident', or rather efficiency, are related – capital itself constitutes an existential risk.

The response from AI enthusiasts, capitalist and socialist, is that AI is precisely what is now needed to avert global warming and other environmental cataclysms. Not only does the very identification of global warming depend on advanced computing infrastructures (Edwards 2010), but the establishment of networks of sensors and monitors measuring and controlling energy, automatically orchestrating an array of clean energy sources, levelling out peak usage, coordinating surge pricing, etc., could, it is claimed, prevent, or at least moderate, climate crisis. AI will exercise a pastoral care over humanity. Such visions of high-technology environmental stewardship are part of the discourse of eco-modernity (Amblee 2018). The difficulty with this attractive vision is that AI, having apparently been converted from being part of the problem to being part of the solution, immediately threatens to turn back into the problem again because AI is a high-energy-use proposition (LePage 2018). Domestic use of electronic gadgets makes a contribution to global warming, but the great data centres crucial to AI are major heat-generating sources. Despite real advances in greening computer technology, some of the most publicized efforts, such as Google's clean data centre policy, are based on offset credits, purchasing heat pollution rights from other companies (Geuss 2018b). As Benjamin Bratton (2016) has described, there is a real possibility that the energy expenditure required for the comprehensive machinic modelling and monitoring of global carbon emissions might actually contribute more to the heating of the planet than it would save. Underlying this problem is the paradox that if AI is being employed simultaneously to promote high economic growth societies, with their vigorous exploitation of the environment, and to mitigate that exploitation, it is running against itself. This, it seems to us, is what is envisaged by left accelerationism and fully automated luxury communism.

Wartime AI

Ecological crisis is not the only vector pushing towards a rethinking of rosy visions of high-tech socialism. In the 'Fragment on Machines', Marx speaks of high-technology, as it existed in his industrial age, exploding the foundations of capitalism. We take this rather literally. The cybernetic origins of AI lie in war, and so may its denouement. The rapidly expanding military application of ML and robotics in so-called New Cold Wars and wars on terror overlies the class-war dynamics we have charted in previous chapters. Discernible in what is widely termed an AI 'arms race' between the US and China are the not-so-faintly inscribed lines of badly fated collisions between declining and ascending great powers. This polarity, however, is only one term in a whole concatenation of shadow and hybrid conflicts waged in part with cybernetic and semi-autonomous weapons systems (Scharre 2018; Dyer-Witheford and Matviyenko 2019).

The twentieth century demonstrated that the only force that can kill capital is capital itself: the proletariat is, *sensu strictu*, just the 'grave-digger'. In 1917 in Russia and 1949 in China, revolution arose from inter-capitalist war. Anything that weak anti-capitalist forces might throw at capital in terms of sabotage, psychological warfare and mischief is dwarfed by what capital is launching against itself in the murky confluence of cyber-war and cyber-crime that render networks increasingly dysfunctional. Nick Land's (2014) compelling vision of an unstoppable AI-capital ascendancy omits the possibility that the competitive dynamics of the world market result in the mutual destruction of contending cybernetic capitals. Given the tight interdependencies of cyber-war and nuclear war, this is a potentially species-fatal dynamic, but it is also potentially a revolution-generating one. The violent fragmentation of the turned-against-itself world market will produce revolts, of very varied political inflection. Communist versions will, as Bernes (2018) observes, inevitably be localized, and, we would add, likely be situated amid rapidly disintegrating networks and degrading infrastructures. This means that, if successful, they will quite possibly take control only of AI in ruins, and that they will do so only at the culmination of struggles in which AI has predominantly been in the hands of those attempting to suppress them.

Communist AI

A communist orientation to AI takes as its priority neither halting AI (Luddism) nor intensifying its development (accelerationism) but rather liquidating the structural dynamics of capital that have so far fostered its development – i.e. capital's imperative to reduce the costs of labour-power as a factor of production and to speed up the circulation of other commodities. Whether or not there might be a 'communist AI' depends on whether 'AI' can exist outside of those conditions. From this point of view, the most promising parts of postcapitalist/left-accelerationist programmes are not those that advance the automation of work within capital, but rather those that point to the expropriation of AI-capital, the development of new forms of collective ownership of AI, and the application of AI to the collectivization of other sectors. At the moment, however, such possibilities appear only at the far end of a spectrum of discussion opened by events such as the Cambridge Analytica–Facebook scandal. Such discussions generally extend from the currently main-streamed prospect of greater self-regulation by AI giants, to intensified government regulation (acceptable in Europe but not North America), to the remote possibility of serious trust-busting action against the concentration of ownership which is a feature of actually-existing AI-capital. On the far edges of this horizon, however, in contrast to capitalist discourse about AI as the new electricity, appear discussions of AI and its related infrastructures as a computational 'public utility' that might actually be subject to democratic control (Mosco 2017). This is the edge that should be pushed forward, towards the deeper concept of a post-revolutionary 'communal utility'. Only where cuts in capital's integument of material intellectual property appear are there prospects for working-class steering of AI development, for the involvement of workers and communities in determining what sorts of work should or should not be automated, and thus for a genuine determination by the 'general intellect' as to the design of AI – in other words, for the revival of aspirations for a digital-era equivalent to the Lucas Plan (the 1970s-era shop-stewards plan for the conversion and remaking of one of the UK's major military-industrial corporations) (Holtwell 2018).

We emphasize the importance of the ongoing conversations and arguments between libertarian commoners and state-oriented socialists in conceiving such a formation (Bauwens and Jose 2018; Murdock 2018). Within such a new form of 'vast association' – to cite Marx's (1848) definition of communism – we can imagine the application of ML to a wide variety of public and communal projects in ways that erode capital's social subsumption and replace it with new priorities. This is where there exist the possibilities of the wide (re)training of AI in logics other than that of the market, which otherwise points to a liquidation of the human as a frictional drag on profit accumulation. But we think that such a true democratization of AI should not be foreclosed by the assumption that its technologies would automatically be accelerated. Working hours might well be substantially reduced by the application of new technologies, with important gains in free time, but postcapitalist 'work' as a site of societal purpose and collective association would not necessarily be annihilated, cutting a swelling 'useless class' loose into an opioids-plus-Netflix wasteland. On the contrary, real collective decision-making about fourth industrial revolution technologies must include the possibility of roads not taken, and humans rejecting the liquidation of their activities by machinic intelligence. This would be a mode of production with something better to say than 'the AI-generated UBI cheque is in the email'.

The trajectory of AI-capital will not be diverted by the reformist measures of AI plus UBI and eco-modern climate planning, on a more-or-less painless path where it automates itself out of existence by increasing its organic composition and pulling the value rug it stands on from under its own feet. While left accelerationism and its allied schools of thought attempt a break with the rendezvous of capital and machine intelligence, they only abet that process.

To find a counter-proposition, we need something else, something akin, though not identical, to what Raniero Panzieri enunciated half a century ago, close to the start of the cybernetic era, when he articulated the doctrine of 'refusal' that lies at the root of what is known as autonomist Marxism: autonomist precisely in seeking, not the autonomy of capital from humans, but of humans from capital:

> [T]he capitalist use of machinery is not, so to speak, a mere distortion of, or deviation from, some 'objective' development that is in itself rational, but that capital has determined technological development ... that 'the science, the gigantic natural forces, and the mass of social labour' are embodied in the system of machinery, which, together with those three forces, constitutes the power of the 'master'. Hence, vis-à-vis the 'voided' individual worker, technological development presents itself as a development of capitalism: as capital, and 'because it is capital, the automatic mechanism is endowed, in the person of the capitalist, with consciousness and a will'. (Panzieri 2017)

Following from this, Panzieri postulated that 'working-class overthrow of the system is a negation of the entire organization in which capitalist development is expressed – and first and foremost of technology insofar as it is linked to productivity' (2017). That said, it is also true that there can be no straightforward recapitulation of *operaismo*'s original 'refusal' strategy, based as it was on the strength of the mass worker in the industrial factory; we are today in a fully social, if not global, factory and in one where the cybernetic processes that Panzieri and his comrades saw coming towards them from over an obscure horizon are now part of the relentless 24/7 light of digital capitalism. In this context, it is quite true that the borders between refusal and reappropriation, between the saboteur, the hacker and the defector, are more fluid and complex than before. If we are sceptical of accelerationism, we also think that Walter Benjamin's famous 'pulling of the emergency brake' is entering the realm of sepia-toned photo-images. The issue at this point is a series of swerves and transversal manoeuvres, as the locomotive of history goes completely off its imaginary tracks. But not all diagonal moves are equal: some lose you the game. All struggles against actually-existing AI-capital are within it, but this does not mean that the moment of negation, rather than acquiescence, can be relinquished: in the twinning of refusal and reappropriation, refusal comes first, if left adoptions of AI are not to be merely accessory to capital.[2]

INHUMAN POWER

A fully developed AI-capitalism would realize the deepest shadows haunting Marxian thought about inhuman power. Chapter 3 explored how we may be progressing towards this future through the emergence of value-producing AGI. This can be understood as a value-theoretic or non-fetishistic analysis of what the original, and very anti-Marxist, accelerationist Nick Land (2014) meant by the 'the teleological identity of capitalism and artificial intelligence'. Although, as we have already stressed, there are many aspects of Land's work with which we are completely unsympathetic, his perspective on AI has the merit of being far franker than that of utopian theorists of the technological singularity such as Kurzweil. For Land (2017), AI is the culmination of a cybernetic process, not in the narrow sense of a particular doctrine of computer development, but in the sense that capital is itself a process of self-reinforcing technological advancement, a 'positive feedback circuit', within which

> commercialization and industrialization mutually excite each other in a runaway process, from which modernity draws its gradient … As the circuit is incrementally closed, or intensified, it exhibits ever greater autonomy, or automation. It becomes more tightly auto-productive (which is only what 'positive feedback' already says). Because it appeals to nothing beyond itself, it is inherently nihilistic. It has no conceivable meaning beside self-amplification. It grows in order to grow. Mankind is its temporary host, not its master. Its only purpose is itself. (Land 2017)

Elsewhere, Land more fully names what is at stake in the emergence of AI when he declares that if such a process is emancipatory, what it emancipates is not a 'human species, who reaches species-being to emancipate human individuals', but only the 'means of production' themselves:

> so in using this word of emancipation, sure, I will totally nod along to it if what is meant by that is capital autonomization … I'm no longer interested in … pretending this is the same thing as what the left really means when they're talking about emancipation. I don't think it is. I think what the left means by emancipation is freedom from capital autonomization. (Vast Abrupt 2018)

Irrespective of the fetishistic nature of Land's analysis, we agree with his conclusions: from our point of view, however, capitalist autonomization is what must be defeated and destroyed.

The great problem with Land's perspective is that the advent of human-free AI-capital is not only hailed as inevitable but greeted with 'adulation' (Goldhill 2017) as a necessary evolutionary supersession. It is this celebratory fatalism about the emergence of inhuman superintelligence that connects Land's writings on AI to his notorious involvement with racist and misogynist 'neoreactionary' strands of the alt-right. Neoreaction (or 'NRx') is an agenda for a futuristic and technological restoration of traditional political hierarchies of race, gender and class, and a resurgence of feudalism where corporate CEOs are the new monarchs. This political current, in which Land's ideas mix with those of figures such as computer-scientist Curtis Yarvin (aka 'Mencius Moldbug'), circulates widely through Silicon Valley and is supported by corporate moguls such as Peter Thiel, forming part of the cultural ambience of AI development (Burrows 2018; Sandifer and Graham 2018).

The confluence between AGI research and Neoreaction is addressed by David Golumbia, whose critique of 'computationalism' (2009) and the attempt of much AI research to detach cognition from embodiment is influenced by and shares many of the same concerns as put forward by feminist and postcolonial theory. Golumbia (2019) argues that the flaws of AI systems in regard to race (and, we would add, gender and class) extend well beyond correctable instances of algorithmic bias. Rather, they lie in the very concept of an abstracted and technologically created 'general intelligence' that mirrors the mindset of a predominantly white (and male) AI research community. In this, he sees affinities between AGI research and the notoriously race-laden search for a measurable and objective 'general IQ'. The quest for AGI is, Golumbia says, the search for a 'Great White Robot God'.[3] His analysis of the white supremacist bias of AI is, however, complicated by the emergence of China as an AI superpower. Nonetheless, Golumbia's argument about the reactionary tendency of attempts to create a 'singularity' that potentially elevates the logic of the dominant social system – racist, sexist and, above all, capitalist – to a level of transcendental authority – is important; he is surely correct to identify a futurological fascist impulse in the affirmative adoption of Land's vision of human-free capitalism by right-wing accelerationist AI developers and computer programmers. In the face of this material instantiation of neoreactionary ideas, it is important to recognize the possibilities Land names, not as a power to embrace, but as a force to oppose. Lands' view of AI-capitalism is a history of Skynet written from the point of view of the Terminator; ours is from the perspective of Sarah Connor.

Capital is already an 'automatic subject', but with AGI it would also become autonomous from the labour of humans and, therefore, humanity. Capitalism could continue, but with inhuman general intelligences representing both sides of the struggle between capital and labour, one side accumulating wealth, while the other continues to work for a wage (whatever form it may take) in machinic misery. All the violent contradictions of capital could continue, but enacted by hyper-intelligent machines. Faced with an AGI that can think faster, do things faster, and which is not bound to a particular morphology with consequent biological needs, such as feeding, breathing and defecation, what would and could human workers do? And what would humans do when inhuman general intelligences started to objectify themselves in the world in a way that slowly or quickly makes the planet less and less habitable for human beings?

Biological corporeality, inefficient and insufficient for valorization at machinic speed, would become an obstacle for capital to overcome: humans would have to discard it to survive. On this issue, AI-capital meets transhumanism, with its aspirations towards the transformation of human biology. Leading intellectuals of actually-existing AI-capital are explicit on this point. Kurzweil proposes that humans must 'transcend [the] limitations of our biological bodies and brains' (2005a: 9). Hans Moravec argues that what really matters in *Homo sapiens* is the mind, the 'rest is merely jelly', and advocates brain emulation – copying the 'identity pattern' and downloading it to hardware – because it enables a disembodied mind to be endowed with 'all the advantages of machines' (1988: 117).

Such speculative scientific proposals for transhumanizing the workforce have flowed into the corporate world. Between 2009 to 2017 Google funded the so-called Singularity University that served as a platform for the ideas of Kurzweil, who also worked for the company on ML and natural language processing (Simonite 2017). Although Google withdrew its support for the institution amidst a series of allegations of shady finances and sexual harassment (McBride 2018), 'Singularity U' has rapidly found other corporate sponsors, including US military-aerospace giant Boeing (Catalano 2018). Meanwhile, Elon Musk, while denouncing the dangers of runaway AI, suggests that the only answer to this threat is to implant computers into human brains so that hominids can cognitively keep pace with their machinic competitors. To this end he has invested in the company Neuralink, a neurotechnology company founded in 2017 to develop brain–computer interfaces: its aim, Musk says, is to achieve 'symbiosis with artificial intelligence' (Hamilton 2018).

To keep up with inhuman AGI labour, humans would have to become equally inhuman, mind and body, as immortal wage-labourers. Together with AGI workers, these no-longer-human beings would make obsolete those who decline transformation. Humans would face a choice: capitalist transhumanism or death. And this choice would generate the ultimate incarnation of Panzieri's refusal to work, precisely because accepting individual death and species extinction would be the only alternative to working for a wage, 24/7, until the heat death of the universe.

Avoiding this choice requires a new mode of production; it therefore entails a communist revolution. But does repudiating the inhumanism of AI-led capitalist development amount then, to a last-ditch defence of classical humanism, a reaffirmation of human exceptionalism and species sovereignty? No. Communism too, should, indeed must, be inhuman. Critique is necessarily enunciated from a human perspective, but de-centring the human from humanity's picture of the universe is both demanded by a scientific worldview and, paradoxically, is a requirement of human species survival. It is necessary to intellectually and viscerally understand humankind as bound-in with and indeed constituted in and by systems of other species and of non-living agencies, including (but not limited to) machines. All modes of production have their own anthropogenesis, thus producing different kinds of human (Read 2017). The 'human' that (possibly) emerges from a struggle against AI-capital will be different from the 'human' that went into it. We see at least two branching paths.

The 'human' in a communist society might reappropriate transhumanism and technologically rework itself. A nascent communist society could either choose or be forced (by ecological collapse, perhaps, or the aftermath of war) to radically modify the physical form of the human, including its cognitive apparatus, metabolic system and body. Marx himself recognized that 'man [*sic*] produces man' (2007: 103). As István Mészáros noted, for Marxists there can never be 'a point in history at which we could say: "now the human substance has been fully realized." For such a fixing would deprive the human being of his [*sic*] essential attribute: his power of "self-mediation" and "self-development"' (1970: 119). While transhumanism has origins in socialist thinkers (Bogdanov 1984; Haldane 1924; Bernal 1969), it came of age in radical libertarian circles and proliferated with little criticism of capitalism (Hughes 2012). If transhumanism is understood, as it defines itself, as an 'intellectual and cultural movement that affirms the possibility and desirability of fundamentally improving the human condition through applied reason, especially by developing and making widely available technologies to eliminate aging and to greatly enhance human intellectual, physical, and psychological capacities' (Humanity+ n.d.), there is, as one of the authors of this book has argued (Steinhoff 2014), nothing inherently anti-communistic about it, despite its longstanding liaison with capital. This perspective is easily associated with Marx's humanist and high modernist moments, for as transhumanists acknowledge, their philosophy 'can be viewed as an extension of humanism, from which it is partially derived' (Humanity+ n.d.). Ongoing transhumanist techno-transformations could, however, raise increasingly difficult problems about the nature of the communality to which communism refers. While our respective stances on these issues differ, we all agree that Marxists should seriously engage with transhumanism, to decouple it from its blindly capitalist trajectory, reflect on Marx's own high modernist tendencies, and delineate a social project to embrace or escape.

The alternative form of communist 'inhumanism' is ecological. To struggle for human autonomy from capital is also to struggle for a recognition of the ecological and cosmic human enmeshments and imbrications that capital obscures and obliterates: to recognize that, in actuality, 'we have never been autonomous' (Nelson and Braun 2017). It deposes the fixity of the human by attending to the species' dependence on, and imbrication in, other living systems, rather than re-centring analysis and politics upon the machines some humans have created to dominate other humans and the natural world. In this regard, capital's AI gambit is perhaps human, all too human: communism must play otherwise. A powerful impetus to this line of thought has come from currents of 'ecosocialism' developed by thinkers such as John Bellamy Foster (2002) and Jason Moore (2015), with the latter more strongly representing the inhuman flattening of ontological distinctions between humanity and nature we think is necessary to challenge capital's tendency to a machinic autonomization. We believe, to varying degrees, that these perspectives can be deepened by linkage to Jane Bennett's analysis of the 'vital materiality that runs through and across bodies, both human and nonhuman' (2010: 178) and Timothy Morton's provocative project to 'turn up the volume of the nonhumans within Marxism' (2017: 61). Such positions are 'inhuman' not, as with transhumanism, through embrace of technology, but rather in the sense pioneered by posthuman feminists (Braidotti 2018), in epistemologically and practically overthrowing both gendered and racialized definitions of humanity and the dominative concept of hominid supremacy over ecological networks.[4]

These intellectual and political currents provide resources for an ecologically oriented departure from the trajectory of automatic AI-capital, forming a line of flight and fight that thinks of human relations to nature, the nonhuman and other humans not in terms of domination or competition, but of mutualism and cooperation. It may be out of the combination of climate change, war and the slow erosion of wage-labour that there will appear the lineaments of a mode of production that, while it must initially formally subsume capital's already-existing AI, goes on to really subsume it into entirely new technological configurations in which AI automation is of far less importance than the cultivation of collective and individual human habits, subjectivities, associations and practices necessary for ecological and social sustainability and species survival. In this respect, one of the most promising directions is that of a possible articulation between Marxism and 'de-growth' movements, such as those adumbrated in recent works by the value-theorist Anselm Jappe (2017) and the autonomist Emmanuelle Leonardi (2019). It could be in such intentionally decelerationist movements that the deep reservoirs of a 'biocommunist' alternative to AI-capital are established.

AI-capital is an abyss, communism a bridge across, but a perilous, shaky one, partially in flames, and with an obscured arrival point on the other side: nonetheless, advance.

Notes

INTRODUCTION

1. One of us (Kjøsen 2013a), without prior knowledge of Land's argument, also arrived at the same conclusions through a discussion of whether androids could labour and create surplus-value.
2. For more on Land and his influence on academia, Silicon Valley and alt-right/neo-fascist/neo-reactionary thought, see Wark (2017), Goldhill (2017) and Haider (2017).
3. Throughout this book we use AI to refer to AI as a whole, including ML. Where the specific properties of ML are important to the discussion at hand we make specific reference to ML. While in most cases the existing AI systems we discuss are ML, this is not always the case. How or whether machine learning is differentiated from statistics are disputed topics we leave to experts. Regardless of one's perspective, there are substantial areas of overlap between the two fields (Srivastava 2015; Fawcett and Hardin 2017).
4. The point deserves emphasis: what is called learning in machines is not the same as learning in humans. Nor does machine perception, or any kind of machine cognition, function the same way that analogous processes or functions do in humans.
5. Marx divides constant capital into 'fixed' and 'circulating'. The former refers to buildings, tools and machines, whereas the latter refers to raw material and other inputs into the production process. Marx distinguishes between the two based on how they transfer their value to the commodity during the production process: either piecemeal (fixed capital) or wholesale (circulating) (see Marx 1992: 237–48).

CHAPTER 1

1. This is far from a complete survey of the commercial possibilities of ML. Indeed, its applications seem endless because ML's ability to find patterns can be applied in any data rich field, which in the digital age is almost any field at all (Zilis and Cham 2016).
2. Srnicek (2016) suggests that Apple, because of its narrow and tightly controlled focus on its high-design computing products, does not really qualify as a 'platform capitalist' dependent on user data, but this assessment can be contested, given both its involvement with music streaming and the app-fuelled iPhone.

3. For a recent review of various attempts to complete Marx's thought on the state, see O'Kane (2014), and for discussion of the state's role in creating and maintaining the general conditions of production, Läpple (1973).
4. However, the very fact that the privatization of initially state-led technologies is so advanced in the US creates some problems for the US government. The big-tech five, Google, Amazon, Facebook, Apple and Microsoft together spend almost as much on non-defence-related scientific research as the US federal government (Manjoo 2017). In instances where there is conflict between the US government and tech companies, as, for example, between the FBI and Apple over encryption, it is by no means certain that the state holds superior know-how.
5. When Marx discussed the particular conditions of production, he referred only to those elements of production that assumed the form of constant capital, and he gave special attention to fixed capital in particular. While this book focuses primarily on the form of fixed capital, it is important to remember that the economic form of fixed capital cannot be reduced to machinery. Buildings, tools, internal transportation networks, and facilities for storing the productive reserve and finished commodities all function as fixed capital in the same way as machinery does (Marx 1990: 510; 1992: 201).
6. Even the ostensibly human internet is increasingly travelled by machines. According to Zeifman (2017), humans only accounted for 48.2 per cent of traffic on the internet in 2016 with the rest being generated by various bots.
7. *Post-operaismo* thinkers have generally not directly addressed the AI issue, with the exception of Matteo Pasquinelli who has elaborated the necessity of understanding information machines and data, especially metadata, as key components of labour and capital in twenty-first-century capitalism and recognizes that capital and AI have deep and evolving affinities. His work represents an attempt to come to terms with some of the contradictions AI presents to *post-operaismo* thought, though he has yet to present a systematic critique. We therefore look forward eagerly to his upcoming monograph on capital and computation (2019).

CHAPTER 2

1. For a sample, see the papers from the conference 'The Economics of Artificial Intelligence: An Agenda', 13–14 September 2017, available from the National Bureau of Economic Research at https://www.nber.org/books/agra-1, published in Agrawal, Gans and Goldfarb (2019).
2. Leo Impett (2018), however, argues that removing algorithmic bias from Deep Learning (DL) AI is all but impossible, as such systems automatically vectorize any and all data; in GOFAI and normal ML, it is possible to remove biases like gender, race, credit score, etc.; DL algorithms take everything into account – which is also why they are so powerful.

CHAPTER 3

1. Discussion of 'the singularity' was popularized by Raymond Kurzweil (2005a). Competing accounts adopt different markers for its arrival. Marxian discussions of the singularity are rare, but include Rikowski (2003); Dyer-Witheford (2010); Kjøsen (2013a; 2018); Rectenwald (2013); Steinhoff (2014).
2. This chapter is an expansion and critique of the arguments one of us presented in the article 'Träumen Androiden vom Mehrwert?' (Kjøsen 2018).
3. HLMI may also refer to 'high-level machine intelligence' (Nadin 2018: 2) or can be understood as 'Human-Level AI (HLAI) (Nilsson 2005).
4. Each mode of production has its own definition of human (Read 2017). That the definition of HMLI ties human intelligence to labour strongly suggests that anthropogenesis in the capitalist mode of production concerns a being's capacity to labour. Indeed, Nils J. Nilsson has suggested that the Turing test be replaced by the 'employment test', which an AI would pass if it is 'able to perform the jobs ordinarily performed by humans' (2005: 68).
5. As active AGI projects, Baum includes projects that 'either identify as AGI or conduct R&D to build something that is considered to be AGI, human-level intelligence, or superintelligence' (2017: 13). Baum includes projects aimed at 'brain emulations' because they would be 'computational entities with general intelligence' and are as such 'a type of AGI' (2017: 8). There is, however, disagreement as to whether brain emulation falls under the umbrella of AGI (Wang and Goertzel 2007: 7). Baum's survey is based on publicly available information which means that there are likely more active AGI projects that have yet to be made publicly known or are deliberately kept in secret.
6. Baum's (2017: 18) breakdown is as follows: academic institutions (20); private corporations (12); public corporations (6); non-profits (5); governmental; no formal institution (2). Four of the projects are split across two of these categories, which is why these numbers add up to 49.
7. As we discuss later, technical achievement is of course by no means the same as economic deployment. For an attempt to estimate the temporal horizons of the technological singularity by an eminent mainstream economist, putting it at least over a century away, if ever, see Nordhaus (2015).
8. Marx referred to this fetish as the 'trinity formula' (1991: 953).
9. As raw material inputs, animals would still appear as constant capital, but of the circulating variety and, therefore, with a different social function than that of fixed (constant) capital.
10. In Moore and Aveling's translation of *Capital*, both *Kopf* and *ideell* are translated as 'imagination' (Marx 2011: 198). While Ben Fowkes (Marx 1990) correctly translates *ideell*, the correct translation of *Kopf* is 'head'.
11. Intuition can be linked to creativity, although the connection is tenuous (Pétervári, Osman and Bhattacharya 2016).

12. CNNs are specific types of neural networks that are modelled on the visual cortex and well-suited to processing image data.
13. They therefore argue that the specific architecture of ANNs is significant to solving the generalization problem and needs to incorporate lateral and feedback connections rather than just resembling the feed-forward hierarchy of cells in the neocortex.
14. ImageNet is a visual database consisting of over 1 million labelled images and 20,000 categories. It was designed for research on visual object recognition software research.
15. BAYOU, which we discussed in Chapter 2, is yet another example.
16. Marx points out that 'machinery is most valuable for capital when its value = 0' (1993: 766).
17. The legal framework suggested by the Committee on Legal Affairs, however, is to 'ensure that robots are and will remain in the service of humans'.
18. This bundle of goods may even be cheaper than the bundle human beings need to reproduce their labour-power. Thus AGI proletarians may emerge as a new pool of very cheap labour because AGI workers would spend a shorter part of their working day on reproducing their wage, i.e. performing necessary labour relative to surplus labour.
19. For a lucid and sympathetic treatment by a non-Marxist economist of Marx's 'surplus population' thesis in relation to AI see Skidelsky (2018).

CONCLUSION

1. This, we note, is the view of AI held some of China's Marxist scholars who believe their government is still on the road to socialism, and that under its tutelage AI will spell the end of capitalism: 'If AI rationally allocates resources through big data analysis, and if robust feedback loops can supplant the imperfections of "the invisible hand" while fairly sharing the vast wealth it creates, a planned economy that actually works could at last be achievable' (Xiang 2018).
2. The same can be said about a predictable attack: Luddism. As far as we are concerned the hour of Luddism has come and gone: the phrase is exhausted. The historical conditions that generated the terms and made them a signifier of seriously threatening or promising subversion have changed so much that neither the adoption nor the damnation of Luddism is of much relevance. All struggles are in and against the conditions of actually-existing AI-capitalism. In terms of destroying digital infrastructures or rendering them uninhabitable by humans, capital is doing the job itself in cyber-wars and cyber-crimes, not to mention eco-disasters, all of which however spur the development of technologies. The question is not the human sabotage of capitalist machinery, but capital's machinic sabotage of humanity, and the possibility of a different social trajectory.
3. The philosopher Reza Negarestani has recently argued for a conception of AGI as the ultimate, emancipatory goal of philosophy and takes issue with both sceptical accounts such as Golumbia's and laudatory ones such

as Land's, holding that 'the project of artificial general intelligence ... [is] a natural extension of the human's process of self-discovery through which the last vestiges of essentialism are washed away' (2018: 117).

4. Such 'posthuman feminism' stands in a complex relation to the accelerationist 'xenofeminism' (Laboria Cuboniks 2015; Hester 2018) that, building on the legacy of socialist-feminist Shulameth Firestone (1979), looks enthusiastically to cybernetic and biotechnologies as means for the 'abolition of gender', and can be seen as a feminist and queer version of transhumanism. The question of whether the radical anti-naturalism of xenofeminism can be reconciled with the ecological concerns of other posthuman feminisms is at this point unresolved. In this regard, the controversies surrounding Donna Haraway's re-modulation of her famous invocation of 'cyborg' radicalism (1985) to her more recent posthuman 'humus' (i.e. earth) based politics (2016) are of particular importance, but also go beyond the scope of our study of AI.

Bibliography

Aarts, Emillie and José Luis Encarnação (2006) *True Visions: The Emergence of Ambient Intelligence*, New York: Springer.

Abdelrahman, Ogail (2017) 'Deep Learning in Production and Warehousing With Amazon Robotics', *Medium*, 2 May, https://medium.com/@teamrework/deep-learning-in-production-warehousing-with-amazon-robotics-571e69fea721

Accenture (2017) 'Accenture Technology Vision 2017: AMPLIFYYOU', www.accenture.com/us-en/insight-artificial-intelligence-ui

Acemoglu, Daron and Pascual Resteropo (2018) 'Artificial Intelligence, Automation and Work', National Bureau of Economic Research, www.nbr.org/papers/w24196

Agrawal, Ajay, Joshua S. Gans and Avi Goldfarb (2018) *Prediction Machines: The Simple Economics of Artificial Intelligence*, Boston: Harvard Business Review Press.

Agrawal, Ajay K., Joshua Gans and Avi Goldfarb (eds) (2019) *The Economics of Artificial Intelligence: An Agenda*, Chicago: University of Chicago Press.

Ajunwa, Ifeoma, Kate Crawford and Jason Schultz (2017) 'Limitless Worker Surveillance', *California Law Review* 105:3.

Alba, Davey (2017) 'The Hidden Laborers Training AI to Keep Ads Off Hateful YouTube Videos', *Wired*, 21 April, www.wired.com/2017/04/zerochaos-google-ads-quality-raters

Alpaydin, Ethem (2016) *Machine Learning: The New AI*, Cambridge MA: MIT Press.

Amadeo, Ron (2018) 'Google's Iron Grip on Android: Controlling Open Source by any Means Necessary', *Ars Technica*, https://arstechnica.com/gadgets/2018/07/googles-iron-grip-on-android-controlling-open-source-by-any-means-necessary

Amazon (2016) 'Introducing Amazon Go and the World's Most Advanced Shopping Technology', www.youtube.com/watch?v=NrmMk1Myrxc

Amazon Workers and Supporters (2018) '"Stop Treating Us Like Dogs!" Workers Organizing Resistance at Amazon in Poland', in Jake Alimahomed-Wilson and Immanuel Ness (eds) *Choke Points: Logistics Workers Disrupting the Global Supply Chain*, London: Pluto.

Amblee, R.S. (2018) *The Ugly Fight: Unleashing Artificial Intelligence Against Global Warming*, New York: Gloture Books.

Amin, Ash (1994) 'Post-Fordism: Models, Fantasies and Phantoms of Transition', in Amin (ed.), *Post-Fordism: A Reader*, Oxford: Wiley-Blackwell.

Anderson, Berit and Brett Horvath (2017) 'The Rise of the Weaponized AI Propaganda Machine', *Scout.ai*, 9 February, https://scout.ai/story/the-rise-of-the-weaponized-ai-propaganda-machine

Arakelyan, Sophia (2017) 'Tech Giants Are Using Open Source Frameworks to Dominate the AI Community', *Venture Beat*, 29 November, https://venturebeat.com/2017/11/29/tech-giants-are-using-open-source-frameworks-to-dominate-the-ai-community

Arntz, Melanie, Terry Gregory and Ulrich Zierahn (2016) 'The Risk of Automation for Jobs in OECD Countries: A Comparative Analysis', OECD Social, Employment and Migration Working Papers 189, http://dx.doi.org/10.1787/5jlz9h56dvq7-en

Asay, Matt (2017) 'Open Source Innovation is Now All About Vendor On-ramps', *InfoWorld*, 30 November, www.infoworld.com/article/3238491/open-source-tools/open-source-innovation-is-now-all-about-vendor-on-ramps.html

Ashton, Kevin (2009) 'That "Internet of Things" Thing', *RFID Journal*, 22:7.

Autor, David (2015) 'Why Are There Still So Many Jobs? The History and Future of Workplace Automation', *Journal of Economic Perspectives* 2:3.

Autor, David H. and David Dor (2013) 'The Growth of Low-Skill Service Jobs and the Polarization of the US Labor Market', *American Economic Review* 103:5.

Babkin, Petr, Md Faisal Mahbub Chowdhury, Alfio Gliozzo, Martin Hirzel and Avraham Shinnar (2017) 'Bootstrapping Chatbots for Novel Domains', Workshop at NIPS on Learning with Limited Labeled Data (LLD), http://hirzels.com/martin/papers/lld17-swagger-nlu.pdf

Bahro, Rudolf (1978) *The Alternative in Eastern Europe*, London: Verso.

Bailly, Christian (2003) *Automata: The Golden Age 1848–1914*, London: Robert Hale.

Banks, Ian M. (1988) *Consider Phlebas*, London: Orbit.

Banks, Ian M. (2013) *The Hydrogen Sonata*, London: Orbit.

Baraniuk, Chris (2018) 'How Talking Machines Are Taking Call Centre Jobs', *BBC News*, 24 August, www.bbc.com/news/business-452728352018

Barth, Brian (2018) 'The Fight Against Google's Smart City', *Washington Post*, 8 August, www.washingtonpost.com/news/theworldpost/wp/2018/08/08/sidewalk-labs/?noredirect=on&utm_term=.1aa977c59316

Bastani, Aaron (2014) 'Fully Automated Luxury Communism', *Novarra Media*, 10 July, https://novaramedia.com/2014/11/10/imo-w-aaron-bastani-e003

Bastani, Aaron (forthcoming 2019) *Fully Automated Luxury Communism: A Manifesto*, New York: Verso.

Baum, Seth D. (2017) 'A Survey of Artificial General Intelligence Projects for Ethics, Risk, and Policy', Global Catastrophic Risk Institute Working Paper 17-1.

Baum, Seth (2018a) 'Superintelligence Skepticism as a Political Tool', *Information* 9.

Baum, Seth (2018b) 'Countering Superintelligence Misinformation', *Information* 9.

Bauwens, Michel and Jose Ramos (2018) 'Re-imagining the Left Through an Ecology of the Commons', *Global Discourse* 8:2.

BBC (2018a) 'Google Drops $10bn Battle for Pentagon Data Contract', *BBC News*, 9 October, www.bbc.com/news/technology-45798153

BBC (2018b) 'Amazon Scrapped "Sexist AI" Tool', *BBC News*, 10 October, www.bbc.com/news/technology-45809919

Beaudry, Paul, David A. Green and Ben Sand (2013) 'The Great Reversal in the Demand for Skill and Cognitive Tasks', NBER Working Paper No. 18901, www.economics.ubc.ca/files/2013/05/pdf_paper_paul-beaudry-great-reversal.pdf

Beer, Randall D. (2014) 'Dynamical Systems and Embedded Cognition', in Keith Frankish and William M. Ram (eds), *The Cambridge Handbook of Artificial Intelligence*, Cambridge: Cambridge University Press.

Benanav, Aaron (2017) 'Automation and the Future of Work', www.versobooks.com/blogs/3412-audio-aaron-benanav-automation-and-the-future-of-work

Benanav, Aaron and John Clegg (2014) 'Misery and Debt: On the Logic and History of Surplus Populations and Surplus Capital', in Andrew Pendakis (ed.), *Contemporary Marxist Theory: An Anthology*, New York: Continuum.

Bennett, Jane (2010) *Vibrant Matter: A Political Ecology of Things*, Durham NC: Duke University Press.

Benton, Ted (1988) 'Humanism = Speciesism: Marx on Humans and Animals', *Radical Philosophy* 50:3.

Benton, Ted (1993) *Natural Relations: Ecology, Animal Rights and Social Justice*, London: Verso.

Benton, Ted (2003) 'Marxism and the Moral Status of Animals', *Society and Animals* 11:1.

Berg, Janine et al. (2018) 'Digital Labour Platforms and the Future of Work: Towards Decent Work in the Online World', International Labour Organisation, www.ilo.org/wcmsp5/groups/public/---dgreports/---dcomm/---publ/documents/publication/wcms_645337.pdf

Bernal, J.D. (1969) [1929] *The World, the Flesh & the Devil: An Enquiry into the Future of the Three Enemies of the Rational Soul*, Bloomington: Indiana University Press.

Bernes, Jason (2018) 'The Belly of the Revolution: Agriculture, Energy, and the Future of Communism', in Brent Ryan Bellamy and Jeff Diamanti (eds), *Materialism and the Critique of Energy*, Chicago: MCM' Publishing.

Bernes, Jasper (2013) 'Logistics, Counterlogistics and the Communist Prospect', *Endnotes* 3, https://endnotes.org.uk/issues/3/en/jasper-bernes-logistics-counterlogistics-and-the-communist-prospect

Bhattacarii, Alex (2017) 'Inside Elon Musk's Disruption Factory', *Wired*, 20 July, www.wired.co.uk/gallery/tesla-factory-fremont-tour-photos-pictures

Bhattacharya, Tithi (ed.) (2018) *Social Reproduction Theory: Remapping Class, Recentering Oppression*, London: Pluto.

Biddle, Sam (2018) 'Facebook Uses Artificial Intelligence to Predict Your Future Actions for Advertisers, Says Confidential Document', *The Intercept*, 13 April, https://theintercept.com/2018/04/13/facebook-advertising-data-artificial-intelligence-ai

Bland, Ben (2016) 'China's Robot Revolution', *Financial Times*, 6 July, www.ft.com/content/1dbd8c60-0cc6-11e6-ad80-67655613c2d6

Bloch, Ernst (1986) [1955] *The Principle of Hope*, Volume 2, Oxford: Oxford University Press.

Boden, Margaret A. (2014) 'GOFAI', in Keith Frankish and William M. Ramsey (eds), *The Cambridge Handbook of Artificial Intelligence*, Cambridge: Cambridge University Press.

Boewe, Jörn and Johannes Schulten (2017) *The Long Struggle of the Amazon Employees: Laboratory of Resistance*, Brussells: Rosa Luxemburg Stiftung.

Bogdanov, Alexander (1984) [1908] *Red Star*, Bloomington: Indiana University Press.

Bost, Matthew W. (2016) 'Entangled Exchange: *Verkehr* and Rhetorical Capitalism', *Review of Communication* 16:4.

Boston Consulting Group–Sutton Trust (2017) *The State of Social Mobility in the UK*.

Bostrom, Nick (2014) *Superintelligence: Paths, Dangers, Strategies*, Oxford: Oxford University Press.

Braga, A. and R.K. Logan (2017) 'The Emperor of Strong AI Has No Clothes: Limits to Artificial Intelligence', *Information* 8:4.

Braidotti, Rosi (2018) 'Posthuman Feminist Theory', in Lisa Disch and Mary Hawkesworth (eds), *The Oxford Handbook of Feminist Theory*, Oxford: Oxford University Press.

Bratton, Benjamin H. (2015) 'Outing A.I.: Beyond the Turing Test', *New York Times*, 23 February, https://opinionator.blogs.nytimes.com/2015/02/23/outing-a-i-beyond-the-turing-test

Bratton, Benjamin H. (2016) *The Stack: On Software and Sovereignty*, Cambridge MA: MIT Press.

Brophy, Enda (2017) *Language Put to Work: The Making of the Global Call Centre Workforce*, New York: Palgrave Macmillan.

Brunton, Finn and Helen Nissenbaum (2015) *Obfuscation: A User's Guide for Privacy and Protest*, Cambridge MA: MIT Press.

Brynjolfsson, Eric and Andrew McAfee (2014) *The Second Machine Age: Work, Progress, and Prosperity in a Time of Brilliant Technologies*, New York: Norton.

Brynjolfsson, Erik and Andrew McAfee (2017) 'The Business of Artificial Intelligence', *Harvard Business Review*, https://hbr.org/cover-story/2017/07/the-business-of-artificial-intelligence

Brynjolfsson, Erik, Daniel Rock and Chad Syverson (2017) 'Artificial Intelligence and the Modern Productivity Paradox: A Clash of Expectations and Statistics', NBER Working Paper No. 24001, www.nber.org/papers/w24001

Buolamwini, Joy (2018) 'InCoding–In The Beginning', *MIT Media Lab*, https://medium.com/mit-media-lab/incoding-in-the-beginning-4e2a5c51a45d

Buranyi, Stephan (2018) '"Dehumanising, Impenetrable, Frustrating": The Grim Reality of Job Hunting in the Age of AI', *Guardian*, 4 March, www.theguardian.

com/inequality/2018/mar/04/dehumanising-impenetrable-frustrating-the-grim-reality-of-job-hunting-in-the-age-of-ai

Burrows, Roger (2018) 'On Neoreaction', *The Sociological Review*, 29 September, www.thesociologicalreview.com/blog/on-neoreaction.html

Byrnes, Nanette (2017) 'As Goldman Embraces Automation, Even the Masters of the Universe Are Threatened', *MIT Technology Review*, 7 February, www.technologyreview.com/s/603431/as-goldman-embraces-automation-even-the-masters-of-the-universe-are-threatened

Caddell, Bud (2017) 'Your Boss Might Be Better As An Algorithm', *Quartz*, 16 November, https://qz.com/1130095/your-boss-might-be-better-as-an-algorithm

Caffentzis, George (2008) 'From the Grundrisse to Capital and Beyond: Then and Now', *Workplace* 15.

Caffentzis, George (2013) *In Letters of Fire and Blood: Work, Machines and the Crisis of Capitalism*, Oakland, CA: PM Press.

Campbell, Peter (2018) 'Trucks Headed for a Driverless Future', *Financial Times*, 30 January, www.ft.com/content/7686ea3e-e0dd-11e7-a0d4–0944c5f49e46

Cant, Callum (2017) 'Precarious Couriers are Leading the Struggle Against Platform Capitalism', *PoliticalCritique.org*, 3 August, http://politicalcritique.org/world/2017/precarious-couriers-are-leading-the-struggle-against-platform-capitalism/#

Čapek, Karel (2004) *R.U.R. (Rossum's Universal Robots)*, London: Penguin.

Catalano, Frank (2018) 'Boeing Helps Lead New $32M Investment in Singularity University, Explores Deeper Partners', *GeekWire*, 15 February, www.geekwire.com/2018/boeing-helps-lead-new-32m-investment-singularity-university-explores-deeper-partnership

Clarke, John (2017) 'Basic Income: Progressive Dreams Meet Neoliberal Realities', *The Bullet* 1350, https://socialistproject.ca/2017/01/b1350

Chan, Jenny (2017) 'Robots, Not Humans: Official Policy in China', *New Internationalist*, 1 November, https://newint.org/features/2017/11/01/industrial-robots-china

Chang, Jae-Hee and Phu Huynh (2016) 'ASEAN Transformation: The Future of Jobs at Risk of Automation', Bureau for Employers' Activities, Working Paper No. 9.

Chua, Charmaine S. (2017) 'Logistical Violence, Logistical Vulnerabilities: A Review of *The Deadly Life of Logistics: Mapping Violence in Global Trade* by Deborah Cowen', *Historical Materialism* 25:4.

Cillo, Rossana and Lucia Pradella (2017) 'Strike Friday at Amazon.it' *Jacobin*, 29 November, www.jacobinmag.com/2017/11/strike-friday-amazon-italy-unions-logistics

Cleaver, Harry (1979) *Reading Capital Politically*, Brighton: Harvester.

Cocchia, Annalisa (2014) 'Smart and Digital City: A Systematic Literature Review', in Renta Paola Dameri and Camille Rosenthal-Sabroux (eds), *Smart City: How to Create Public and Economic Value With High Technology in Urban Space*, New York: Springer.

Cockshott, Paul and Karen Renaud (2016) 'Humans, Robots and Values', *Technology in Society* 45.

Coeckelbergh, Mark (2017) 'Can Machines Create Art?', *Philosophy & Technology* 30:3.

Cohen, Nicole and Greig de Peuter (2018) 'I Work at Vice Canada and I Need a Union', in Stephanie Ross and Larry Savage (eds), *Labour Under Attack: Anti-Unionism in Canada*, Winnipeg: Fernwood.

Coles, Terri (2018) 'How AI Trading Systems Will Shake Up Wall Street', *IT Pro Today*, 12 January, www.itprotoday.com/machine-learning/how-ai-trading-systems-will-shake-wall-street

Collins, Harry (2018) *Artifictional Intelligence: Against Humanity's Surrender to Computers*, Cambridge: Polity.

Columbus, Louis (2016) '10 Ways Machine Learning is Revolutionizing Manufacturing', *Forbes*, 26 June, www.forbes.com/sites/louiscolumbus/2016/06/26/10-ways-machine-learning-is-revolutionizing-manufacturing/#4f8f7eff28c2

Cook, Mike (2018) 'A Basic Lack of Understanding', *Notes From Below*, 3 March, http://notesfrombelow.org/article/a-basic-lack-of-understanding

Copeland, Jack (2000) 'What is Artificial Intelligence?', *AlanTuring.net*, www.alanturing.net/turing_archive/pages/reference%20articles/what_is_AI/What%20is%20AI11.html

Corbet, Jonathan and Greg Kroah-Hartman (2018) '2017 Linux Kernel Development Report', *Linux Foundation*, www.linuxfoundation.org/2017-linux-kernel-report-landing-page

Costanza-Chock, Sasha (2018) 'Design Justice, A.I., and Escape from the Matrix of Domination', *The Journal of Design and Science*, 17 July, https://jods.mitpress.mit.edu/pub/costanza-chock?version=8cf4cf92-96d4-4d09-92ac-8865b4dcfc07

Cowen, Deborah (2014) *The Deadly Life of Logistics: Mapping Violence in Global Trade*, Minneapolis: University of Minnesota Press.

Crosby, Simon (2018) 'Open-Source Machine Learning is Free, as in Beer', *Forbes*, www.forbes.com/sites/forbestechcouncil/2018/10/04/open-source-machine-learning-is-free-as-in-beer/#17b2b1e943d1

Cuppini, Niccolò, Mattia Frapporti and Maurilio Pirone (2015) 'Logistics Struggles in the Po Valley Region: Territorial Transformations and Processes of Antagonistic Subjectivation', *South Atlantic Quarterly* 114:1.

Dale, Robert (2016) 'The Return of the Chatbots', *Natural Language Engineering* 22:5.

De Stefano, Valerie (2018) '"Negotiating the Algorithm": Automation, Artificial Intelligence and the Protection of Labour', International Labour Organisation Working Paper 246, www.ilo.org/employment/Whatwedo/Publications/working-papers/WCMS_634157/lang--en/index.htm

DeBord, Matthew (2017) 'Tesla's Future is Completely Inhuman – And We Shouldn't Be Surprised', *Business Insider*, 20 May, www.businessinsider.com/tesla-completely-inhuman-automated-factory-2017-5

Degenerate Communism (2014) 'Chokepoints: Mapping an Anticapitalist Counter-Logistics in California', *Libcom.org*, https://libcom.org/library/chokepoints-mapping-anticapitalist-counter-logistics-california

Del Ray, Jason (2017) 'Amazon's Store of the Future Has No Cashiers, But Humans are Watching From Behind the Scenes', *Recode*, www.recode.net/2017/1/6/14189880/amazon-go-convenience-store-computer-vision-humans

Despoudis, Fanis (2017) 'How Machine Learning and AI Could Eventually Replace Development Work', *Codeburst*, 11 September, https://codeburst.io/how-machine-learning-and-ai-could-eventually-replace-development-work-922ebfod59co

Dettmers, Tim (2015) 'Deep Learning in a Nutshell: Core Concepts', *NVidia Developer Blog*, https://devblogs.nvidia.com/deep-learning-nutshell-core-concepts

Dinerstein, Ana Cecilia and Frederick Harry Pitts (2018) 'From Post-Work to Post-Capitalism? Discussing the Basic Income and Struggles for Alternative Forms of Social Reproduction', *Journal of Labor and Society*, https://onlinelibrary.wiley.com/journal/24714607

Domingos, Pedro (2015) *The Master Algorithm: How the Quest for the Ultimate Learning Machine Will Remake Our World*, New York: Basic Books.

Dong, Catherine (2017) 'FBLearner Flow: The Evolution of Machine Learning', *TechCrunch*, 8 August, https://techcrunch.com/2017/08/08/the-evolution-of-machine-learning

Drake, P. (2015) 'Marxism and the Nonhuman Turn: Animating Nonhumans, Exploitation and Politics with ANT and Animal Studies', *Rethinking Marxism* 27:1.

Drexler, Eric K. (1987) *Engines of Creation: The Coming Era of Nanotechnology*, New York: Anchor Library of Science.

Dreyfus, Hubert (1972) *What Computers Can't Do*, Cambridge MA: MIT Press.

Dutton, Tim (2018) 'An Overview of National AI Strategies', *Politics + AI Blog*, https://medium.com/politics-ai/an-overview-of-national-ai-strategies-2a70ec6edfd

Dyer-Witheford, Nick (1999) *Cyber-Marx: Cycles and Circuits of Struggle in High-technology Capitalism*, Urbana: University of Illinois Press.

Dyer-Witheford, Nick (2010) 'Digital Labour, Species-becoming and the Global Worker', *Ephemera* 10:3/4.

Dyer-Witheford, Nick (2014) 'Red Plenty Platforms', *Culture Machine* 14.

Dyer-Witheford, Nick (2015) *Cyber-proletariat: Global Labour in the Digital Vortex*, London: Pluto.

Dyer-Witheford, Nick and Svitlana Matviyenko (2019) *Cyberwar and Revolution: Digital Subterfuge in Global Capitalism*, Minneapolis: University of Minnesota Press.

Economist (2013) 'Workers' Share of National Income: Labour Pains', *The Economist*, 31 October, www.economist.com/finance-and-economics/2013/10/31/labour-pains

Economist (2016a) 'Artificial Intelligence: The Return of the Machinery Question', 25 June, www.economist.com/news/special-report/21700758-will-smarter-machines-cause-mass-unemployment-automation-and-anxiety

Economist (2016b) 'Rise of the Superstars', 17 Sep, www.economist.com/news/special-report/21707048-small-group-giant-companiessome-old-some-neware-once-again-dominating-global

Economist (2017a) 'The Human Cumulus', *The Economist*, 26 August, www.economist.com/business/2017/08/26/artificial-intelligence-will-create-new-kinds-of-work

Economist (2017b) 'Battle of the Brains: Google Leads in the Race to Dominate Artificial Intelligence', *The Economist*, 7 December, www.economist.com/business/2017/12/07/google-leads-in-the-race-to-dominate-artificial-intelligence

Economist (2017c) 'Learning and Earning: Special Report-Lifelong Education', 14 January, www.economist.com/news/special-report/21714169-technological-change-demands-stronger-and-more-continuous-connections-between-education

Economist (2018) 'Economists Grapple with the Future of the Labour Market', *The Economist*, 11 January, www.economist.com/finance-and-economics/2018/01/11/economists-grapple-with-the-future-of-the-labour-market

Eden, David (2012) *Autonomy: Capitalism, Class and Politics*, Farnham: Ashgate.

Edwards, Paul (1996) *The Closed World: Computers and the Politics of Discourse in Cold War America*, Cambridge MA: MIT Press.

Edwards, Paul (2010) *A Vast Machine: Computer Models, Climate Data, and the Politics of Global Warming*, Cambridge MA: MIT Press.

Elliott, Larry (2017) 'Robots Will Not Lead to Fewer Jobs – But the Hollowing Out of the Middle Class', *Guardian*, 20 August, www.theguardian.com/business/2017/aug/20/robots-are-not-destroying-jobs-but-they-are-hollow-out-the-middle-class

Elster, Jon (1985) *Making Sense of Marx*, Cambridge: Cambridge University Press.

Endnotes (2010) 'The History of Subsumption', *Endnotes* 2, https://endnotes.org.uk/issues/2/en/endnotes-the-history-of-subsumption

Ensign, Danielle, Sorelle A. Friedler, Scott Neville, Carlos Scheidegger and Suresh Venkatasubramanian (2017) 'Runaway Feedback Loops in Predictive Policing', *arXiv.org*, https://arxiv.org/abs/1706.09847

Ertel, Wolfgang (2018) *Introduction to Artificial Intelligence*, New York: Springer.

Eubanks, Virginia (2017) *Automating Inequality: How High-Tech Tools Profile, Police and Punish the Poor*, New York: St. Martin's Press.

Fabian (2018) 'Global Artificial Intelligence Landscape', *Medium*, https://medium.com/@bootstrappingme/global-artificial-intelligence-landscape-including-database-with-3-465-ai-companies-3bf01a175c5d

Faggella, Daniel (2018a) 'Crowdsourced Sentiment Analysis – Applications in Social Media and Customer Service', *TechEmergence*, www.techemergence.com/crowdsourced-sentiment-analysis-applications-social-media-customer-service

Faggella, Daniel (2018b) 'Industrial AI Applications – How Time Series and Sensor Data Improve Processes', *TechEmergence*, www.techemergence.com/industrial-ai-applications-time-series-sensor-data-improve-processes

Faggella, Daniel (2018c) 'What is Artificial Intelligence? An Informed Definition', *Emerj*, https://emerj.com/ai-glossary-terms/what-is-artificial-intelligence-an-informed-definition

Fawcett, Tom and Drew Hardin (2017) 'Machine Learning vs. Statistics', *Silicon Valley Data Science*, www.svds.com/machine-learning-vs-statistics

Federici, Silvia (2012) *Revolution at Point Zero: Housework, Reproduction, and Feminist Struggle*, New York: PM Press.

Financial Stability Board (FSB) (2017) 'Artificial Intelligence and Machine Learning in Financial Services: Market Developments and Financial Stability Implications', *Financial Stability Board*, www.fsb.org/wp-content/uploads/P011117.pdf

Finley, Klint (2016) 'Microsoft Open-sources its Artificial Brain to One-up Google', *Wired*, www.wired.com/2016/01/microsoft-tries-to-one-up-google-in-the-open-source-ai-race

Firestone, Shulamith (1970) *The Dialectic of Sex: The Case for Feminist Revolution*, New York: William Morrow.

Foer, Franklin (2017) *World Without Mind: The Existential Threat of Big Tech*, New York: Penguin.

Ford, Martin (2009) *The Lights in the Tunnel: Automation, Accelerating Technology and the Economy of the Future*, New York: Acculant Publishing.

Ford, Martin (2016) *Rise of the Robots: Technology and the Threat of a Jobless Future*, New York: Basic Books.

Foster, John Bellamy (2002) *Ecology Against Capitalism*, New York: Monthly Review Press.

Fracchia, J. (2017) 'Organisms and Objectifications: A Historical-Materialist Inquiry into the "Human and Animal"', *Monthly Review* 68:10.

Frank, Morgan R., Lijun Sun, Manuel Cebrian, Hyejin Youn and Iyad Rahwan (2018) 'Small Cities Face Greater Impact from Automation', *Journal of the Royal Society Interface* 15:139.

Frenkel, Shara (2018) 'Microsoft Employees Protest Work With ICE, as Tech Industry Mobilizes Over Immigration', *New York Times*, 19 June, www.nytimes.com/2018/06/19/technology/tech-companies-immigration-border.html

Frey, Carl Benedict and Michael A. Osborne (2013) 'The Future of Employment: How Susceptible are Jobs To Computerisation?', www.oxfordmartin.ox.ac.uk/downloads/academic/The_Future_of_Employment.pdf

Future of Life Institute (2018) 'AI Policy – United States', https://futureoflife.org/ai-policy-united-states

Gagné, Jean-François (2018) 'Global AI Talent Report 2018', www.jfgagne.ai/talent

Gallagher, Sean (2013a) 'What the NSA Can Do With "Big Data"', *Ars Technica*, 11 June, https://arstechnica.com/information-technology/2013/06/what-the-nsa-can-do-with-big-data

Gallagher, Sean (2013b) 'You May Already Be a Winner in NSA's "Three-Degrees" Surveillance Sweepstakes!', 18 July, https://arstechnica.com/information-technology/2013/07/you-may-already-be-a-winner-in-nsas-three-degrees-surveillance-sweepstakes

Gallagher, Sean (2018) 'The Snowden Legacy, Part One: What's Changed, Really?', *Ars Technica*, 21 November, https://arstechnica.com/tech-policy/2018/11/the-snowden-legacy-part-one-whats-changed-really

Gao, Jack (2017) 'China's Wage Growth: How Fast is the Gain and What Does It Mean?', *New Institute for Economic Thinking*, www.ineteconomics.org/perspectives/blog/chinas-wage-growth-how-fast-is-the-gain-and-what-does-it-mean

Gartner (n.d.) 'Knowledge Capture', *Gartner IT Glossary*, www.gartner.com/it-glossary/knowledge-capture

Gartner (2017) 'Gartner Says By 2020, Artificial Intelligence Will Create More Jobs Than it Eliminates', 13 December, www.gartner.com/newsroom/id/3837763

Gent, Ed (2017) 'The Hidden Human Workforce Powering Artificial Intelligence', *Singularity Hub*, https://singularityhub.com/2017/12/19/the-hidden-human-workforce-powering-machine-intelligence/#sm.0001d2qlh93ildkhvt12nmij5pv6c

Gent, Ed (2018) 'The Democratization of AI is Putting Powerful Tools in the Hands of Non-Experts', *Singularity Hub*, https://singularityhub.com/2018/02/19/the-democratization-of-ai-is-putting-powerful-tools-in-the-hands-of-non-experts/#sm.0001d2qlh93ildkhvt12nmij5pv6c

George, Dileep et al. (2017) 'A Generative Vision Model that Trains With High Data Efficiency and Breaks Text-Based CAPTCHAs', *Science* 358:1271.

Gershgorn, Dave (2015) 'How Google Aims to Dominate Artificial Intelligence', *Popular Science*, 9 November, www.popsci.com/google-ai

Gershgorn, Dave (2018) 'Google Gave the World Powerful AI Tools, and the World Made Porn with Them', *Quartz*, https://qz.com/1199850/google-gave-the-world-powerful-open-source-ai-tools-and-the-world-made-porn-with-them

Geuss, Megan (2018a) 'Low Pay, Poor Prospects, and Psychological Toll: The Perils of Microtask Work', *Ars Technica*, 23 September, https://arstechnica.com/information-technology/2018/09/in-most-cases-online-microtask-work-can-be-a-raw-deal-un-study-finds

Geuss, Martin (2018b) 'Google Wants to Match Actual Energy Demand with Carbon-free Supply', *Ars Technica*, 11 October, https://arstechnica.com/information-technology/2018/10/googles-data-center-carbon-heat-maps-show-the-challenges-of-going-carbon-free

Ghaffrey, Shirin (2018) 'Many in Silicon Valley Support Universal Basic Income. Now the California Democratic Party Does, Too', *Recode*, 8 March, www.recode.net/2018/3/8/17081618/tech-solution-economic-inequality-universal-basic-income-part-democratic-party-platform-california

Gibbs, Samuel (2015) 'Women Less Likely to be Shown Ads for High-paid Jobs on Google, Study Shows', *Guardian*, 8 July, www.theguardian.com/technology/2015/jul/08/women-less-likely-ads-high-paid-jobs-google-study

Gillespie, Tarleton (2010) 'The Politics of "Platforms"', *New Media and Society* 12:3.

Goertzel, Ben (2007) 'Human-level Artificial General Intelligence and the Possibility of a Technological Singularity. A Reaction to Ray Kurzweil's *The Singularity Is Near*, and McDermott's Critique of Kurzweil', *Artificial Intelligence* 171.

Goertzel, Ben (2014) 'Artificial General Intelligence: Concept, State of the Art, and Future Prospects', *Journal of Artificial General Intelligence* 5:1.

Goldhill, Olivia (2017) 'The Neo-fascist Philosophy that Underpins Both the Alt-right and Silicon Valley Technophiles', *Quartz*, 18 June, https://qz.com/1007144/the-neo-fascist-philosophy-that-underpins-both-the-alt-right-and-silicon-valley-technophiles

Golumbia, David (2009) *The Cultural Logic of Computation*, Cambridge MA: Harvard University Press.

Golumbia, David (2019) 'The Great White Robot God: Artificial General Intelligence and White Supremacy', *Uncomputing*, 21 January, www.uncomputing.org/?p=2077

Goode, Eric and Claire Cain Miller (2013) 'Backlash by the Bay: Tech Riches Alter a City', *New York Times*, 24 November, www.nytimes.com/2013/11/25/us/backlash-by-the-bay-tech-riches-alter-a-city.html?_r=0

Goodfellow, Ian et al. (2014) 'Generative Adversarial Nets', *Proceedings of the 27th International Conference on Neural Information Processing Systems* 2.

Gordon, Robert (2016) *The Rise and Fall of American Growth: The U.S. Standard of Living Since the Civil War*, Princeton: Princeton University Press.

Gosaduaff, Laurence (2017) '2018 Will Mark the Beginning of AI Democratization', *Smarter with Gartner*, www.gartner.com/smarterwithgartner/2018-will-mark-the-beginning-of-ai-democratization

Gourevitch, Alex and Lucas Stanczyk (2018) 'The Basic Income Illusion', *Catalyst* 1:4.

Grace, Katja, John Salvatier, Allan Dafoe, Baobao Zhang and Owain Evans (2018) 'Viewpoint: When Will AI Exceed Human Performance? Evidence from AI Experts', *Journal of Artificial Intelligence Research* 62.

Gray, Mary and Siddharth Suri (2017) 'The Humans Working Behind the AI Curtain', *Harvard Business Review*, 9 January, https://hbr.org/2017/01/the-humans-working-behind-the-ai-curtain

Greene, Tristan (2018) 'AI Guru Fei Fei Li Set to Leave Google this Year', *The Next Web*, https://thenextweb.com/artificial-intelligence/2018/09/10/ai-guru-fei-fei-li-set-to-leave-google-this-year

Greenfield, Adam (2013) *Against the Smart City*, New York: Do Projects.

Greenmeier, Larry (2008) 'Artificial Intelligence: Robots Rule When it Comes to Holiday Shopping', *Scientific American*, 26 December, www.scientificamerican.com/article/artificial-intelligence-robots-rule

Griffiths, Thomas L., Nick Chater, Charles Kemp, Amy Perfors and Joshua B. Tenenbaum (2010) 'Probabilistic Models of Cognition: Exploring Representations and Inductive Biases', *Trends in Cognitive Sciences* 14:8.

Grothoth, Christian and J.M. Porup (2016) 'The NSA's SKYNET Program May be Killing Thousands of Innocent People', *Ars Technica*, 1 February, https://arstechnica.com/information-technology/2016/02/the-nsas-skynet-program-may-be-killing-thousands-of-innocent-people

Gubrud, M. (1997) 'Nanotechnology and International Security', *Fifth Foresight Conference on Molecular Nanotechnology* 1, https://foresight.org/Conferences/MNT05/Papers/Gubrud

Gunn, Richard (1987) 'Marxism and Mediation', *Common Sense* 2.

Haider, Shuja (2017) 'The Darkness at the End of the Tunnel: Artificial Intelligence and Neoreaction', *Viewpoint Magazine*, 28 March, www.viewpointmag.com/2017/03/28/the-darkness-at-the-end-of-the-tunnel-artificial-intelligence-and-neoreaction

Haldane, J.B.S. (1924) *Daedalus, or, Science and the Future*, London: K. Paul, Trench, Trubner.

Halliday, James (2018) 'Open Source is Not Enough', *Notes From Below* 2, https://notesfrombelow.org/article/open-source-is-not-enough

Hamilton, Isobel Asher (2018) 'Elon Musk Believes AI Could Turn Humans into an Endangered Species Like the Mountain Gorilla', *Business Insider*, 26 November, www.businessinsider.com/elon-musk-ai-could-turn-humans-into-endangered-species-2018–11

Harari, Yuval Noah (2016) *Homo Deus: A Brief History of Tomorrow*, New York: Signal Press.

Haraway, Donna (1985) 'A Manifesto for Cyborgs: Science, Technology, and Socialist Feminism in the 1980s', *Socialist Review* 80.

Haraway, Donna (2016) 'Staying With the Trouble: Anthropocene, Capitalocene, Chthulucene', in Jason Moore (ed.), *Anthropocene or Capitalocene? Nature, History and the Crisis of Capitalism*, Oakland, CA: PM Press.

Hardt, Michael and Antonio Negri (2001) *Empire*, Cambridge MA: Harvard University Press.

Hardt, Michael and Antonio Negri (2017) *Assembly*, Oxford: Oxford University Press.

Harris, Mark (2014) 'Amazon's Mechanical Turk Workers Protest: "I am a Human Being, Not an Algorithm"', *Guardian*, 3 December, www.theguardian.com/technology/2014/dec/03/amazon-mechanical-turk-workers-protest-jeff-bezos

Harvey, Cynthia (2017) 'Open Source Artificial Intelligence: 50 Top Projects', *Datamation*, www.datamation.com/open-source/open-source-artificial-intelligence-50-top-projects-1.html

Hawksworth, John, Richard Berriman and Saloni Goel (2018) 'Will Robots Really Steal our Jobs? An International Analysis of the Potential Long Term Impact of Automation', PricewaterhouseCoopers, www.pwc.co.uk/economic-services/assets/international-impact-of-automation-feb-2018.pdf

He, Kaiming, Xiangyu Zhang, Shaoqing Ren and Jian Sun (2016) 'Deep Residual Learning for Image Recognition', *Proceedings of the IEEE Conference on Computer Vision and Pattern Recognition.*

Heinrich, Michael (2012) *An Introduction to the Three Volumes of Karl Marx's Capital*, New York: Monthly Review Press.

Henwood, Doug (2015) 'Glum Job Prospects, Say Officials', *LBO News*, 21 December, https://lbo-news.com/2015/12/21/glum-job-prospects-say-officials

Henwood, Doug (2018a) 'About that Stock Panic', *Jacobin*, 2 February, https://jacobinmag.com/2018/02/stock-market-drop-wage-increases-economy

Henwood, Doug (2018b) 'The Gig Economy Fantasy', *Jacobin*, 15 June, www.jacobinmag.com/2018/06/precarity-american-workplace-gig-economy

Hermann, Jeremy and Mike Del Baso (2017) 'Meet Michelangelo: Uber's Machine Learning Platform', *Uber Engineering*, 5 September, https://eng.uber.com/michelangelo.

Hern, Alex (2017) 'Give Robots "Personhood" Status, EU Committee Argues', *Guardian*, 12 January, www.theguardian.com/technology/2017/jan/12/give-robots-personhood-status-eu-committee-argues

Hernandez, Pedro (2017) 'Artificial Intelligence to Assimilate Software Industry', *Datamation*, www.datamation.com/applications/artificial-intelligence-to-assimilate-software-industry.html

Hernandez, Dan (2018) '"Robots Can't Beat us": Las Vegas Casino Workers Prep for Strike Over Automation', *Guardian*, 2 June, www.theguardian.com/us-news/2018/jun/02/las-vegas-workers-strike-automation-casinos

Hester, Helen (2016) 'Technically Female: Women, Machines, and Hyperemployment', *Salvage*, 8 August, http://salvage.zone/in-print/technically-female-women-machines-and-hyperemployment

Hester, Helen (2018) *Xenofeminism*, Cambridge: Polity.

Hill, Steven (2015) *Raw Deal: How the 'Uber Economy' and Runaway Capitalism are Screwing American Workers*, New York: St. Martin's Press.

Holtwell, Felix (2018) 'Bringing Back the Lucas Plan', *Notes From Below*, https://notesfrombelow.org/article/bringing-back-the-lucas-plan

Hook, Leslie (2016) 'The Humans Behind Mechanical Turk's Artificial Intelligence', *Financial Times*, 26 October, www.ft.com/content/17518034-6f77-11e6-9ac1-1055824ca907

Horwitz, Josh (2018) 'China is Both Ahead of and Behind Amazon in Cashier-less Stores', *Quartz*, 22 January, https://qz.com/1185081/amazon-go-china-is-both-ahead-of-and-behind-amazon-in-cashier-less-stores

Howcroft, Debra and Jill Rubery (2018) 'Gender Equality Prospects in the Fourth Industrial Revolution', in Max Neufeind et al. (eds), *Work in the Digital Age: Challenges of the Fourth Industrial Revolution*, New York: Roman and Littlefield.

Hughes, Chris (2018) *Fair Shot: Rethinking Inequality and How We Earn*, New York: St. Martin's Press.

Hughes, James J. (2012) 'The Politics of Transhumanism and the Techno-Millennial Imagination, 1626–2030', *Zygon* 47:4

Humanity+ (n.d.) 'Transhumanist FAQ', https://humanityplus.org/philosophy/transhumanist-faq

Huws, Ursula (2014) *Labor in the Global Digital Economy: The Cybertariat Comes of Age*, New York: Monthly Review.

Ilyas, Ihab (2018) 'Data Cleaning is a Machine Learning Problem that Needs Data Systems Help', *ACM Sigmod Blog*, 18 April, http://wp.sigmod.org

Impett, Leo (2018) 'Marxism and New Developments in Science: Artificial Intelligence and Deep Learning: Technical and Political Challenges', *Theory & Struggle* 119.

International Data Corporation (2016) 'Press Release: IDC Sees the Dawn of the DX Economy and the Rise of the Digital-Native Enterprise', www.idc.com/getdoc.jsp?containerId=prUS41888916

International Data Corporation (2018) 'Worldwide Spending on Cognitive and Artificial Intelligence Systems Will Grow to $19.1 Billion in 2018, According to New IDC Spending Guide', www.idc.com/getdoc.jsp?containerId=prUS43662418

International Transport Forum (ITF) (2017) 'Managing the Transition to Driverless Road Freight Transport', *International Transport Forum*, www.itf-oecd.org/managing-transition-driverless-road-freight-transport

Ito, Joi (2018) 'The Paradox of Universal Basic Income', *Wired*, 29 March, www.wired.com/story/the-paradox-of-universal-basic-income

Jameson, Fredric (1991) *Postmodernism, or, The Cultural Logic of Late Capitalism*, London: Verso.

Jang, Eric (2017) 'What Companies are Winning the Race for Artificial Intelligence?', *Forbes*, 24 February, www.forbes.com/sites/quora/2017/02/24/what-companies-are-winning-the-race-for-artificial-intelligence/#4902bfe8f5cd

Jappe, Anselm (2017) 'Degrowthers, One More Effort if You Want to be Revolutionaries', in *The Writing on the Wall: On the Decomposition of Capital and Its Critics*, Alresford: Zero Books.

Jenkins, Simon (2018) 'Worrying About Robots Stealing Our Jobs? How Silly', *Guardian*, 20 August, www.theguardian.com/commentisfree/2018/aug/20/robots-stealing-jobs-digital-age

Johnson, E.R. (2017) 'At the Limits of Species Being: Sensing the Anthropocene', *The South Atlantic Quarterly* 116:2.

Johnson, Matthew, Katja Hofmann, Tim Hutton and David Bignell (2016) 'The Malmo Platform for Artificial Intelligence Experimentation', *Proceedings of the Twenty-Fifth International Joint Conference on Artificial Intelligence* ((I) JCAI-16), www.ijcai.org/Proceedings/16/Papers/643.pdf

Johnston, David (2015) 'Let Data Scientists be Data Mungers', *Thought Works*, 5 August, www.thoughtworks.com/insights/blog/let-data-scientists-be-data-mungers

Johnston, John (2008) *The Allure of Machinic Life: Cybernetics, Artificial Life, and the New AI*, Cambridge MA: MIT Press.

Jordan, Tim (2015) *Information Politics: Liberation and Exploitation in the Digital Society*, London: Pluto.

Kale, Sirin (2018) 'Logged Off: Meet the Teens Who Refuse to Use Social Media', *Guardian*, 29 August, www.theguardian.com/society/2018/aug/29/teens-desert-social-media

Kallis, Giorgos and Swyngedouw, Erik (2018) 'Do Bees Produce Value? A Conversation Between an Ecological Economist and a Marxist Geographer', *Capitalism, Nature, Socialism* 29:3.

Kaplan, Jerry (2016) *Artificial Intelligence: What Everyone Needs to Know*, Oxford: Oxford University Press.

Kaput, Mike (2018) 'How the European Union's GDPR Rules Impact Artificial Intelligence and Machine Learning', *Marketing Artificial Intelligence Institute*, 24 May, www.marketingaiinstitute.com/blog/how-the-european-unions-gdpr-rules-impact-artificial-intelligence-and-machine-learning

Katz, Miranda (2017) 'Amazon's Turker Crowd Has Had Enough', *Wired*, 23 August, www.wired.com/story/amazons-turker-crowd-has-had-enough

Kelly, Amy (2018) 'Robot Workers Will Lead to Surge in Slavery in South-East Asia, Report Finds', *Guardian*, 12 July, www.theguardian.com/global-development/2018/jul/12/robot-workers-will-lead-to-surge-in-slavery-in-south-east-asia-report-finds

Kelly, Kevin (2014) 'The Three Breakthroughs That Have Finally Unleashed AI on the World', *Wired*, 27 September, www.wired.com/2014/10/future-of-artificial-intelligence

Kessler, Sarah (2018) *Gigged: The End of the Job and the Future of Work*, New York: St. Martin's Press.

Kinder, Molly (2018) 'Learning to Work With Robots: AI Will Change Everything. Workers Must Adapt – Or Else', *Foreign Affairs*, 11 July, https://foreignpolicy.com/2018/07/11/learning-to-work-with-robots-automation-ai-labor

Kisner, James, David Wishnow and Timur Ivannikov (2017) 'IBM: Creating Shareholder Value with AI? Not So Elementary, My Dear Watson', *Jeffries*, 12 July, https://javatar.bluematrix.com/pdf/fO5xcWjc

Kitchin, Rob (2014) 'The Real-time City? Big Data and Smart Urbanism', *GeoJournal* 79:1.

Kjøsen, Atle Mikkola (2013a) 'Do Androids Dream of Surplus Value?', Conference paper, Mediations 2.5, London, Ontario, 18 January, www.academia.edu/2455476/Do_Androids_Dream_of_Surplus_Value

Kjøsen, Atle Mikkola (2013b) 'Human Material in the Communication of Capital,' *communication +1*, 2:3.

Kjøsen, Atle Mikkola (2016) *Capital's Media: The Physical Conditions of Circulation*, PhD thesis, University of Western Ontario.

Kjøsen, Atle Mikkola (2018) 'Träumen Androiden vom Mehrwert?', *Maske und Kothurn* 64:1–2.

Knight, Will (2016a) 'Inside Vicarious, the Secretive AI Startup Bringing Imagination to Computers', *MIT Technology Review*, www.technologyreview.com/s/601496/inside-vicarious-the-secretive-ai-startup-bringing-imagination-to-computers

Knight, Will (2016b) 'Machines Can Now Recognize Something After Seeing It Once', *MIT Technology Review*, www.technologyreview.com/s/602779/machines-can-now-recognize-something-after-seeing-it-once

Knight, Will (2017) 'Reinforcement Learning', *MIT Technology Review*, www.technologyreview.com/s/603501/10-breakthrough-technologies-2017-reinforcement-learning

Kofman, Ava (2018) 'Google's "Smart City of Surveillance" Faces New Resistance in Toronto', *The Intercept*, 13 November, https://theintercept.com/2018/11/13/google-quayside-toronto-smart-city.

Kolanovic, Marko and Rajesh T. Krishnamachari (2017) 'Big Data and AI Strategies: Machine Learning and Alternative Data Approach to Investing', J.P. Morgan, www.cfasociety.org/cleveland/Lists/Events%20Calendar/Attachments/1045/BIG-Data_AI-JPMmay2017.pdf

Kolchin, Peter (1993) *American Slavery 1619–1877*, New York: Hill & Wang.

Kolinko (2002) *Hotlines - call centre | inquiry | communism*, www.nadir.org/nadir/initiativ/kolinko/lebuk/e_lebuk.htm

Kopetz, Hermann (2011) *Real-time Systems: Design Principles for Distributed Embedded Applications*, Boston: Springer.

Kulian, Artur (2017) 'Why Decentralized Artificial Intelligence Will Reinvent the Industry as We Know It', *Forbes*, 16 November, www.forbes.com/sites/forbestechcouncil/2017/11/16/why-decentralized-artificial-intelligence-will-reinvent-the-industry-as-we-know-it/#f7d393511a44

Kulka, Benjamin and Richard Brown (2018) 'Human Capital: Disruption, Opportunity and Resilience in London's Workforce', *Centre for London*, 12 August, www.centreforlondon.org/wp-content/uploads/2018/04/Centre-for-London_Human-Capital-Digital-Report.pdf

Kurzweil, Ray (2005a) *The Singularity Is Near*, New York: Penguin.

Kurzweil, Ray (2005b) 'Long Live AI', *Forbes*, www.forbes.com/home/free_forbes/2005/0815/030.html

Kyung-Hoon, Kim (2018) 'How Robots Could Help Care for Japan's Aging Population', *Independent*, 9 April, www.independent.co.uk/arts-entertainment/photography/japan-robot-elderly-care-ageing-population-exercises-movement-a8295706.html

Laboria Cuboniks (2015) 'Xenofeminism: A Politics for Alienation', www.laboriacuboniks.net/index.html#zero/1

Laird, J.E. (2012) *The Soar Cognitive Architecture*, Cambridge MA: MIT Press.

Lake, Brendan. M., Tomer D. Ullman, Joshua B. Tenenbaum and Samuel J. Gershman (2017) 'Building Machines That Learn and Think Like People', *Behavioral and Brain Sciences* 40.

Land, Nick (2012) *Fanged Noumena: Collected Writings 1987–2007*, Falmouth: Urbanomic.

Land, Nick (2014) 'The Teleological Identity of Capitalism and Artificial Intelligence: Remarks to the Participants of the Incredible Machines 2014 Conference', 8 March, formerly available at http://incrediblemachines.info/nick-land-the-teleological-identity-of-capitalism-and-artificial-intelligence

Land, Nick (2017) 'A Quick and Dirty Introduction to Accelerationism', *Jacobite*, https://jacobitemag.com/2017/05/25/a-quick-and-dirty-introduction-to-accelerationism

Landing A.I. (n.d.) 'Transform Your Business With A.I.', www.landing.ai

Lapovsky, Issie (2014) 'The Next Big Thing You Missed: The Quest to Give Computers the Power of Imagination', *Wired*, www.wired.com/2014/04/vicarious-ai-imagination

Läpple, Dieter (1973) *Staat und allgemeine Produktionsbedingungen. Grundlagen zur Kritik der Infrastrukturtheorien*, Berlin: Verlag für das Studium der Arbeiterbewegung.

Larson, Christina (2018) 'Closing the Factory Doors', *Foreign Affairs*, 16 July, https://foreignpolicy.com/2018/07/16/closing-the-factory-doors-manufacturing-economy-automation-jobs-developing

Larson, Jeff, Surya Mattu, Lauren Kirchner and Julia Angwin (2016) 'How We Analyzed the COMPAS Recidivism Algorithm', *ProPublica*, 23 May, www.newscientist.com/article/2166207-discriminating-algorithms-5-times-ai-showed-prejudice

Lebeuf, Carlene, Margaret-Anne Storey and Alexey Zagalsky (2017) 'How Software Developers Mitigate Collaboration Friction with Chatbots', arXiv preprint:1702.07011

LeCun, Yann, Yoshua Bengio and Geoffrey Hinton (2015) 'Deep Learning', *Nature* 521:7553.

Lee, Kai-Fu (2018) *AI Superpowers: China, Silicon Valley, and the New World Order*, New York: Houghton-Mifflin.

Lee, Timothy (2018) 'Waymo Jobs: Self-driving Cars Will Destroy a Lot of Jobs – They'll Also Create a Lot', *Ars Technica*, 24 August, https://arstechnica.com/tech-policy/2018/08/self-driving-cars-will-destroy-a-lot-of-jobs-theyll-also-create-a-lot

Lenat, D.B. and J.S. Brown (1984) 'Why AM and EURISKO Appear to Work', *Artificial Intelligence* 23:3.

Leonardi, Emmanuele (2019) 'Bringing Class Analysis Back in: Assessing the Transformation of the Value-Nature Nexus to Strengthen the Connection Between Degrowth and Environmental Justice', *Ecological Economics* 156.

LePage, Michael (2018) 'AI's Dirty Secret: Energy-guzzling Machines May Fuel Global Warming', *New Scientist*, 10 October, www.newscientist.com/article/mg24031992–100-ais-dirty-secret-energy-guzzling-machines-may-fuel-global-warming

Levy, Frank (2018) 'Computers and Populism: Artificial Intelligence, Jobs, and Politics in the Near Term', *Oxford Review of Economic Policy* 34:3.

Levy, Steven (2018) 'Inside Amazon's Artificial Intelligence Flywheel', *Wired*, 2 January, www.wired.com/story/amazon-artificial-intelligence-flywheel

Lewis, Leo (2017) 'Can Robots Make up for Japan's Care Home Shortfall?', *Financial Times*, 17 October, www.ft.com/content/418ffd08–9e10–11e7–8b50–0b9f565a23e1

Leviathan, Yaniv and Yossi Matias (2018) 'Google Duplex: An AI System for Accomplishing Real-World Tasks Over the Phone', *Google AI Blog*, 8 May, https://ai.googleblog.com/2018/05/duplex-ai-system-for-natural-conversation.html

Liu, Wendy (2018) 'Freedom isn't Free', *Logic*, https://logicmag.io/05-freedom-isnt-free

Locke, Charles (2017) 'Void Star: Terrifying Silicon Valley Sci-Fi Only an AI Expert Could Pen', *Wired*, 11 April, www.wired.com/2017/04/void-star-dystopia-computational-linguistics

Lohr, Steve (2014) 'For Big Data Scientists, "Janitor Work" is Key Hurdle to Insights', *New York Times*, www.nytimes.com/2014/08/18/technology/for-big-data-scientists-hurdle-to-insights-is-janitor-work.html

Lum, Kristian and William Isaac (2016) 'To Predict and Serve?', *Significance* 13:5, https://rss.onlinelibrary.wiley.com/doi/full/10.1111/j.1740–9713.2016.00960.x

Lynch, Shana (2017) 'Andrew Ng: Why AI is the New Electricity', *Insights by Stanford Business*, www.gsb.stanford.edu/insights/andrew-ng-why-ai-new-electricity

McBride, Sarah (2018) 'Silicon Valley's Singularity University Has Some Serious Reality Problems', *Bloomberg Businessweek*, 15 February, www.bloomberg.com/news/articles/2018–02–15/silicon-valley-s-singularity-university-has-some-serious-reality-problems

McCarthy, John, Martin Minsky, N. Rochester and Claude Shannon (1955) 'A Proposal for the Dartmouth Summer Research Project on Artificial Intelligence', http://www-formal.stanford.edu/jmc/history/dartmouth/dartmouth.html

McCorduck, Pamela (2004) *Machines Who Think*, 2nd edn, Natick MA: A.K. Peters Ltd.

MacLeod, Ken (2008) *Fractions*, London: Orbit.

MacLeod, Ken (2009) *Divisions*, London: Orbit.

Mandel, Ernest (1975) *Late Capitalism*, London: New Left Books.

Mandel, Ernest (1990) 'Introduction to "Results of the Immediate Process of Production"', in Karl Marx, *Capital, Volume I*, Harmondsworth: Penguin.

Manjoo, Farhad (2017) 'How 5 Tech Giants Have Become More Like Governments Than Companies', *NPR*, 26 October, www.npr.org/2017/10/26/560136311/how-5-tech-giants-have-become-more-like-governments-than-companies

Manyika, James, Susan Lund, Michael Chui, Jacques Bughin, Jonathan Woetzel, Parul Batra, Ryan Ko and Saurabh Sanghvi (2017) 'Jobs Lost, Jobs Gained: What the Future of Work Will Mean for Jobs, Skills, and Wages', McKinsey Global Institute, www.mckinsey.com/featured-insights/future-of-work/jobs-lost-jobs-gained-what-the-future-of-work-will-mean-for-jobs-skills-and-wages

Manzerolle, Vincent and Atle Mikkola Kjøsen (2015) 'Digital Media and Capital's Logic of Acceleration', Christian Fuchs (ed.), *Marx in the Age of Digital Capitalism*, Leiden: Brill.

Marr, Bernard (2016) 'How Analytics, Big Data and AI are Changing Call Centers Forever', *Forbes*, 6 September, www.forbes.com/sites/bernardmarr/

2016/09/06/how-analytics-big-data-and-ai-are-changing-call-centers-forever/#253669b13a32

Marr, Bernard (2018) 'Does Synthetic Data Hold the Secret to Artificial Intelligence?', *Forbes*, 5 November, https://preview.tinyurl.com/y7lvv9ss

Marshall, Aarian (2017) 'The Human-Robocar War for Jobs is Finally On', *Wired*, 29 September, www.wired.com/story/trucks-robocar-senate-war

Marx, Karl (1973) *Contribution to a Critique of Political Economy*, Moscow: Progress Publishers.

Marx, Karl (1975) *Early Writings*, Harmondsworth: Penguin.

Marx, Karl (1990) *Capital, Volume I*, Harmondsworth: Penguin.

Marx, Karl (1991) *Capital, Volume III*, Harmondsworth: Penguin.

Marx, Karl (1992) *Capital, Volume II*, Harmondsworth: Penguin.

Marx, Karl (1993) *Grundrisse*, Harmondsworth: Penguin.

Marx, Karl (2007) *Economic and Philosophic Manuscripts of 1844*, Mineola: Dover Publications.

Marx, Karl (2008) *The Poverty of Philosophy*, New York: Cosimo Press.

Marx, Karl (2011) *Capital, Volume I*, trans. S. Moore and E. Aveling, Mineola: Dover Publications.

Mason, Paul (2015) *Postcapitalism: A Guide to Our Future*, New York: Allen Lane.

Mason, Zachary (2017) *Void Star*, New York: Farrar, Straus and Giroux.

Mayer, Jane (2018) 'A Parlor Game at Rebekah Mercer's Has No Get Out of Jail Free Card', *The New Yorker*, 2 July, www.newyorker.com/magazine/2018/07/02/a-parlor-game-at-rebekah-mercers-has-no-get-out-of-jail-free-card

Mazzucato, Mariana (2013) *The Entrepreneurial State: Debunking Public vs. Private Sector Myths*, London: Anthem Press.

Melville, Andrew (2017) 'Amazon Go is About Payments, Not Grocery', *Forbes*, www.forbes.com/sites/groupthink/2017/01/20/amazon-go-is-about-payments-not-grocery/#5019341567e4

Merchant, Brian (2015) 'Fully Automated Luxury Communism', *Guardian*, 18 March, www.theguardian.com/sustainable-business/2015/mar/18/fully-automated-luxury-communism-robots-employment

Mészáros, István (1970) *Marx's Theory of Alienation*, London: Merlin Press.

Metz, Cade (2016) 'In Two Moves Alphago and Lee Sedol Redefined the Future', *Wired*, www.wired.com/2016/03/two-moves-alphago-lee-sedol-redefined-future

Metz, Cade (2017a) 'Tech Giants are Paying Huge Salaries for Scarce A.I. Talent', *New York Times*, 22 October, www.nytimes.com/2017/10/22/technology/artificial-intelligence-experts-salaries.html

Metz, Cade (2017b) 'Building A.I. that Can Build A.I.', *New York Times*, 7 November, www.nytimes.com/2017/11/05/technology/machine-learning-artificial-intelligence-ai.html

Metz, Cade (2018) 'Big Bets on A.I. Open a New Frontier for Chip Start-Ups', *New York Times*, 14 January, www.nytimes.com/2018/01/14/technology/artificial-intelligence-chip-start-ups.html

Metz, Rachel (2018) 'Google Demos Duplex, its AI that Sounds Exactly Like a Very Weird, Nice Human', *MIT Technology Review*, 27 June, www.

technologyreview.com/s/611539/google-demos-duplex-its-ai-that-sounds-exactly-like-a-very-weird-nice-human

Meyer, David (2018) 'AI Has a Big Privacy Problem and Europe's New Data Protection Law is About to Expose It', *Forbes*, 25 May, http://fortune.com/2018/05/25/ai-machine-learning-privacy-gdpr

Microsoft (n.d.) 'Democratizing AI: For Every Person and Every Organization', https://news.microsoft.com/features/democratizing-ai

Middleton, Chris (2018) 'Europe Announces €20 Billion AI Strategy – UK Sidelined', *Internet of Business*, 25 April, https://internetofbusiness.com/european-commission-announces-new-e20-billion-ai-strategy

Miller, Paul (2018) 'What is Edge Computing?', *The Verge*, 7 May, www.theverge.com/circuitbreaker/2018/5/7/17327584/edge-computing-cloud-google-microsoft-apple-amazon

Moody, Kim (2018a) 'High Tech, Low Growth: Robots and the Future of Work', *Historical Materialism* 26:4.

Moody, Kim (2018b) 'Where's the Gig Economy?', *Jacobin*, 23 June, www.jacobinmag.com/2018/06/gig-economy-precarity-jobs-employment.

Moore, Jason W. (2015) *Capitalism in the Web of Life*, London: Verso.

Moore, Phoebe V., Martin Upchurch and Xanthia Whittaker (eds) (2018) *Humans and Machines at Work: Monitoring, Surveillance and Automation in Contemporary Capitalism*, New York: Palgrave.

Moravec, Hans (1988) *Mind Children: The Future of Robot and Human Intelligence*, Cambridge MA: Harvard University Press.

Morozov, Evgeny (2015) 'Socialize the Data Centres!', *New Left Review* 91.

Morrison, Catherine (2018) 'Bank of England Economist Warns Thousands of UK Jobs at Risk from Robots and AI', *Independent*, 18 August, www.independent.co.uk/news/business/news/uk-job-loss-risk-ai-robots-artificial-intelligence-technology-bank-of-england-andy-haldane-a8498901.html

Morrison, Donn, Ruili Wang and Liyanage C. De Silva (2007) 'Ensemble Methods for Spoken Emotion Recognition in Call-centres', *Speech Communication* 49:2.

Morris-Suzuki, Tessa (1984) 'Robots and Capitalism', *New Left Review* 1:47.

Morris-Suzuki, Tessa (1986) 'Capitalism in the Computer Age', *New Left Review*, 1:60.

Morton, Timothy (2017) *Humankind: Solidarity with Nonhuman People*, London: Verso.

Mosco, Vincent (2014) *To the Cloud: Big Data in a Turbulent World*, London: Paradigm.

Mosco, Vincent (2017) *Becoming Digital: Toward a Post-Internet Society*, Bingley: Emerald Publishing.

Moulier-Boutang, Yann (2011) *Cognitive Capitalism*, Cambridge: Polity.

MSV, Janikiram (2018) 'The Rise of Artificial Intelligence as a Service in the Public Cloud', *Forbes*, www.forbes.com/sites/janakirammsv/2018/02/22/the-rise-of-artificial-intelligence-as-a-service-in-the-public-cloud/#3a512198198e

Muehlhauser, Luke (2013) 'What is AGI?', *MIRI*, 11 August, https://intelligence.org/2013/08/11/what-is-agi

Mulhall, Stephen (1998) 'Species-Being, Teleology and Individuality, Part I: Marx on Species-Being', *Angelaki* 3:1.

Müller, Vincent C. and Nick Bostrom (2016) 'Future Progress in Artificial Intelligence: A Survey of Expert Opinion', in Vincent Müller (ed.), *Fundamental Issues of Artificial Intelligence*, Berlin: Springer.

Murdock, Graham (2018) 'Commons Manifestos: A Reply to Bauwens and Ramos', *Global Discourse* 8:2.

Nadin, Mihail (2018) 'Machine Intelligence: A Chimera', *AI and Society*, https://doi.org/10.1007/s00146-018-0842-8

Nakashima, Ryan (2018a) 'AI's Dirty Little Secret: It's Powered by People', *Phys. Org*, 5 March, https://phys.org/news/2018-03-ai-dirty-secret-powered-people.html

Nakashima, Ryan (2018b) 'Google's AI Push Comes with Plenty of People Problems', *CBS Fox WVSNTV*, 1 February, www.wvnstv.com/news/googles-ai-push-comes-with-plenty-of-people-problems/948618199

Nakashima, Ryan (2018c) 'Artificial Intelligence Has Some Unexpected Help: The Intellect of Humans', *Portland Press Herald*, 3 March.

Nedelkoska, Ljubica and Glenda Quintini (2018) 'Automation, Skills Use and Training', OECD Social, Employment and Migration Working Papers No. 202, http://dx.doi.org/10.1787/2e2f4eea-en

Negarestani, Reza (2018) *Intelligence and Spirit*, Falmouth: Urbanomic/Sequence Press.

Negri, Antonio (2017) *Marx and Foucault: Essays*, Cambridge: Polity.

Nelson, Sara and Bruce Braun (2017) 'Autonomia in the Anthropocene: New Challenges to Radical Politics', *South Atlantic Quarterly* 116:2.

Nesbitt, Jeff (2017) 'Google's True Origin Partly Lies in CIA and NSA Research Grants for Mass Surveillance', *Quartz*, 8 December, https://qz.com/1145669/googles-true-origin-partly-lies-in-cia-and-nsa-research-grants-for-mass-surveillance

Ng, Andrew (2015) 'What Data Scientists Should Know About Deep Learning', www.youtube.com/watch?v=O0VNopGgBZM

Nguyen, Anh, Jason Yosinski and Jeff Clune (2015) 'Deep Neural Networks are Easily Fooled: High Confidence Predictions for Unrecognizable Images', in *Computer Vision and Pattern Recognition*, IEEE, www.evolvingai.org/files/DNNsEasilyFooled_cvpr15.pdf

Nichols, Shaun (2017) 'TV Anchor Says Live On-air "Alexa, Order Me a Dollhouse" – Guess What Happens Next', *The Register*, 17 January, www.theregister.co.uk/2017/01/07/tv_anchor_says_alexa_buy_me_a_dollhouse_and_she_does

Nieva, Richard (2018) 'How Facebook Uses Artificial Intelligence to Take Down Abusive Posts', *CNet*, 2 May, www.cnet.com/news/heres-how-facebook-uses-artificial-intelligence-to-take-down-abusive-posts-f8

Nilsson, Nils (2005) 'Human-Level Artificial Intelligence? Be Serious!', *AI Magazine* 26:4.

Noble, Safiya (2018) *Algorithms of Oppression: How Search Engines Reinforce Racism*, New York: New York University Press.

Nordhaus, William D. (2015) 'Are We Approaching an Economic Singularity? Information Technology and the Future of Economic Growth', National Bureau of Economic Research Working Paper 21547, 1 September, www.nber.org/papers/w21547

Notes From Below (2018a) 'The Workers' Inquiry and Social Composition', *Notes From Below* 2, 29 January, https://notesfrombelow.org/article/workers-inquiry-and-social-composition

Notes From Below (2018b) 'Editors' Notes on Class Composition and Technology', *Notes From Below* 2, 30 March, http://notesfrombelow.org/article/editors-notes-on-class-composition-and-technology

OECD (2016) 'Automation and Independent Work in a Digital Economy', Policy Brief on the Future of Work, www.oecd.org/employment/future-of-work.htm

O'Kane, Chris (2014) 'State Violence, State Control: Marxist State Theory and the Critique of Political Economy', *Viewpoint Magazine*, 29 October, www.viewpointmag.com/2014/10/29/state-violence-state-control-marxist-state-theory-and-the-critique-of-political-economy

Omohundro, Stephen M. (2008) 'The Basic AI Drives', *Proceedings of the 2008 Conference on Artificial General Intelligence 2008: Proceedings of the First AGI Conference.*

O'Neill, Cathy (2016) *Weapons of Math Destruction: How Big Data Increases Inequality and Threatens Democracy*, New York: Random House.

OpenAI (2018) 'OpenAI Five', *OpenAI Blog*, https://blog.openai.com/openai-five

Oracle (2016) 'Can Virtual Experiences Replace Reality? The Future Role for Humans in Delivering Customer Experience', www.oracle.com/webfolder/s/delivery_production/docs/FY16h1/doc35/CXResearchVirtualExperiences.pdf

Panetta, Kasey (2016) 'The ABC Technologies Will Change Future Customer Experience', *Smarter with Gartner*, 7 November, www.gartner.com/smarterwithgartner/the-abc-technologies-will-change-future-customer-experience

Panetta, Kasey (2017) 'Gartner Top 10 Strategic Technology Trends for 2018', *Smarter with Gartner*, 3 October, www.gartner.com/smarterwithgartner/gartner-top-10-strategic-technology-trends-for-2018

Pang, Bo and Lillian Lee (2004) 'A Sentimental Education: Sentiment Analysis Using Subjectivity Summarization Based on Minimum Cuts', in *Proceedings of the 42nd Annual Meeting of Association for Computational Linguistics.*

Panzieri, Raniero (2017) [1961] 'The Capitalist Use of Machinery: Marx Versus the Objectivists', *Libcom.org*, https://libcom.org/library/capalist-use-machinery-raniero-panzieri

Paquette, Daniel (2018) 'At Work, With Only Robots for Company', *Toronto Star*, 15 September, reprinted from the Washington Post, www.pressreader.com/canada/toronto-star/20180915/282699048030639

Park, Morgan (2018) 'How the OpenAI Five Tore Apart a Team of Dota2 Pros', *PC Gamer*, www.pcgamer.com/how-the-openai-five-tore-apart-a-team-of-dota-2-pros

Pasquinelli, Matteo (2019) *The Eye of the Master: Capital as Computation and Cognition*, London: Verso.

Patrizio, Andy (2018) 'Top 25 Artificial Intelligence Companies', *Datamation*, 10 April, www.datamation.com/applications/top-25-artificial-intelligence-companies.html

Pearlstein, Steven (2018) 'The Robots-vs.-Robots Trading that Has Hijacked the Stock Market', *Washington Post*, 7 February, www.washingtonpost.com/news/wonk/wp/2018/02/07/the-robots-v-robots-trading-that-has-hijacked-the-stock-market/?utm_term=.31c249c8f820

Pennachin, Cassio and Ben Goertzel (2007) 'Contemporary Approaches to Artificial General Intelligence', in Cassio Pennachin and Ben Goertzel (eds), *Artificial General Intelligence*, Berlin: Springer-Verlag.

Perlo, Katherine (2002) 'Marxism and the Underdog', *Society and Animals* 10:3.

Peters, Michael A., Roderigo Britiz and Ergin Bulut (2009) 'Cybernetic Capitalism, Informationalism and Cognitive Labour', *Geopolitics, History, and International Relations* 1:2.

Peterson, Andrea (2016) 'Uber's Self-driving Truck Delivered 50,000 Cans of Budweiser', *Washington Post*, 25 October, www.washingtonpost.com/news/the-switch/wp/2016/10/25/ubers-self-driving-truck-delivered-50000-cans-of-budweiser/?noredirect=onandutm_term=.99d27d99f308

Pétervári, Judit, Magda Osman and Joydeep Bhattacharya (2016) 'The Role of Intuition in the Generation and Evaluation Stages of Creativity', *Frontiers in Psychology* 7, Article ID 1420.

Pirrone, Gabrielle (2018) 'Walmart and Microsoft Team Up in Joint Interest to Beat Amazon', *CMS Connected*, 24 July, www.cms-connected.com/News-Archive/July-2018/Walmart-Microsoft-Team-Up-in-Joint-Interest-to-Beat-Amazon

Pitts, Frederick Harry (2017) 'Beyond the Fragment: Postoperaismo, Postcapitalism and Marx's "Notes on Machines", 45 Years On', *Economy and Society* 46:3–4.

Pitts, Frederick Harry and Ana Cecilia Dinerstein (2017) 'Corbynism's Conveyor Belt of Ideas: Postcapitalism and the Politics of Social Reproduction', *Capital and Class* 41:3.

Powers, Shawn M. and Michael Jablonski (2015) *The Real Cyber War: The Political Economy of Internet Freedom*, Urbana: University of Illinois Press.

Press, Gil (2018) '45 Numbers to Keep Track of the AI Bubble', *Forbes*, 2 April, www.forbes.com/sites/gilpress/2018/04/02/45-numbers-to-keep-track-of-the-ai-bubble/#3e6299c87a75

PricewaterhouseCoopers (PwC) (2017) 'The Economic Impact of Artificial Intelligence on the UK Economy', www.pwc.co.uk/economic-services/assets/ai-uk-report-v2.pdf

Prindle, Drew (2016) 'This Automated Store in Sweden Doesn't Have Any Human Employees – Only a Smartphone App', *Digital Trends*, 26 February, www.digitaltrends.com/cool-tech/sweden-app-enabled-automated-store

Pudwell, Sam (2017) 'How Deliveroo is Using Big Data and Machine Learning to Power Food Delivery', *Silicon*, www.silicon.co.uk/data-storage/bigdata/deliveroo-big-data-deliveries-216163

Purdy, Mark and Paul Daugherty (2017) 'Why Artificial intelligence is the Future of Growth', *Accenture Canada*, 24 May, www.accenture.com/t20170524T055435__w__/ca-en/_acnmedia/PDF-52/Accenture-Why-AI-is-the-Future-of-Growth.pdf

Rajan, Amol (2018) 'Why Tech is Taking a Hammering', *BBC News*, 23 November, www.bbc.com/news/entertainment-arts-46317205

Ramsay, Anders (2009) 'Marx? Which Marx?', *Eurozine*, www.eurozine.com/marx-which-marx

Ramtin, Ramin (1991) *Capitalism and Automation: Revolution in Technology and Capitalist Breakdown*, London: Pluto.

Rankin, Jennifer (2018) 'Artificial Intelligence: €20bn Investment Call from EU Commission', *Guardian*, 25 April, www.theguardian.com/technology/2018/apr/25/european-commission-ai-artificial-intelligence

Rao, Anand and Gerard Verweij (2018) 'Sizing the Prize: What's the Real Value of AI for Your Business and How Can You Capitalise?', Pricewaterhouse Coopers, www.pwc.com/gx/en/issues/data-and-analytics/publications/artificial-intelligence-study.html.

Rapp, Nicolas and Brian O'Keefe (2018) 'These 100 Companies are Leading the Way in A.I.', *Fortune*, 8 January, http://fortune.com/2018/01/08/artificial-intelligence-ai-companies-invest-startups

Rayo, Edgar Alan (2018) 'AI and AL Adoption Survey Results from Applied Artificial Intelligence Conference 2017', *TechEmergence*, www.techemergence.com/ai-and-ml-adoption-survey-results-from-applied-artificial-intelligence-conference-2017

Rayome, Alison DeNisco (2018) 'Developers, Rejoice: Now AI Can Write Code for You', *TechRepublic*, www.techrepublic.com/article/developers-rejoice-now-ai-can-write-code-for-you

Read, Jason (2017) 'Man is a Werewolf to Man: Capital and the Limits of Political Anthropology', *Continental Thought & Theory* 1:4.

Rectenwald, Michael (2013) 'The Singularity and Socialism', *Insurgent Notes*, 5 October, http://insurgentnotes.com/2013/10/the-singularity-and-socialism

Reese, Ellen and Jason Struana (2018) '"Work Hard, Make History": Oppression and Resistance in Inland Southern California's Warehouse and Distribution Industry', in Jake Alimahomed-Wilson and Immanuel Ness (eds), *Choke Points: Logistics Workers Disrupting the Global Supply Chain*, London: Pluto.

Reese, Hope (2016a) 'How Data and Machine Learning are Part of Uber's DNA', *TechRepublic*, 21 October, www.techrepublic.com/article/how-data-and-machine-learning-are-part-of-ubers-dna

Reese Hope (2016b) 'Bias in Machine Learning, and How to Stop It', *TechRepublic*, 18 November, www.techrepublic.com/article/bias-in-machine-learning-and-how-to-stop-it

Reese, Hope and Nick Heath (2016) 'Inside Amazon's Clickworker Platform: How Half a Million People are Being Paid Pennies to Train AI', *TechRepublic*, 21 December, www.techrepublic.com/article/inside-amazons-clickworker-platform-how-half-a-million-people-are-training-ai-for-pennies-per-task

Reynolds, Matt (2018) 'Biased Policing is Made Worse by Errors in Pre-crime Algorithms', *New Scientist*, 27 April, www.newscientist.com/article/mg23631464-300-biased-policing-is-made-worse-by-errors-in-pre-crime-algorithms

Rich, Elaine (1983) *Artificial Intelligence*, New York: McGraw-Hill.

Rikowski, Glenn (2003), 'Alien Life: Marx and the Future of the Human', *Historical Materialism* 11:2.

Riot Research (2018) 'AI: Show me the Money: Which Sectors Will Make a Killing in AI Forecast 2017–2023: Executive Summary', *Rethink Research*, 19 July, https://rethinkresearch.biz/report/ai-show-me-the-money-executive-summary-free-download

Roberts, Michael (2018) 'The Productivity Paradox Again', *Michael Robert's Blog*, 29 June, https://thenextrecession.wordpress.com/2018/06/29/the-productivity-puzzle-again

Roberts, Sarah T. (2016) 'Commercial Content Moderation: Digital Laborers' Dirty Work', in S.U. Noble and B. Tynes (eds), *The Intersectional Internet: Race, Sex, Class and Culture Online*, New York: Peter Lang.

Roberts, Sarah T. (2017) 'Social Media's Silent Filter', *The Atlantic*, 8 March, www.theatlantic.com/technology/archive/2017/03/commercial-content-moderation/518796

Robins, Kevin and Frank Webster (1988) 'Cybernetic Capitalism: Information, Technology, Everyday Life', in Vincent Mosco and Janet Wasko (eds), *The Political Economy of Information*, Madison: University of Wisconsin Press.

Robinson, Kim Stanley (2013) *2312*, London: Orbit.

Rodrik, Denis (2015) 'Premature Deindustrialization', http://drodrik.scholar.harvard.edu/files/dani-rodrik/files/premature_deindustrialization_revised2.pdf

Roland, Alex and Philip Shiman (2002) *Strategic Computing: DARPA and the Quest for Machine Intelligence, 1983–1993*, Cambridge MA: MIT Press.

Romano-Lax, Andromeda (2018) *Plum Rains*, New York: Penguin.

Rosenblat, Alex (2016) 'The Truth About How Uber's App Manages Drivers', *Harvard Business Review*, 6 April, https://hbr.org/2016/04/the-truth-about-how-ubers-app-manages-drivers

Rosenblat, Alex (2018) *Uberland: How Algorithms are Rewriting the Rules of Work*, Oakland: University of California Press.

Rossiter, Ned (2016) *Software, Infrastructure, Labor: A Media Theory of Logistical Nightmares*, New York: Routledge.

Ritter, Samuel, David Barrett, Antonio Santoro and Matt Botvinick (2017) 'Cognitive Psychology for Deep Neural Networks: A Shape Bias Case Study', *Proceedings of the 34th International Conference on Machine Learning* 70.

Rubin, Isaak Illich (1973) *Essays on Marx's Theory of Value*, Montreal: Black Rose Books.

Rushkoff, Douglas (2018) 'Universal Basic Income is Silicon Valley's Latest Scam', *Medium*, 10 October, https://medium.com/s/free-money/universal-basic-income-is-silicon-valleys-latest-scam-fd3e130b69a0

Ruth, Joao-Pierre (2017) '6 Examples of AI in Business Intelligence Applications', *TechEmergence*, 8 May, www.techemergence.com/ai-in-business-intelligence-applications

Sadowski, Jathan (2018) 'Why Silicon Valley is Embracing Universal Basic Income', *Guardian*, 2 June, www.theguardian.com/technology/2016/jun/22/silicon-valley-universal-basic-income-y-combinator

Sadowski, Jathan and Frank A. Pasquale (2015) 'The Spectrum of Control: A Social Theory of the Smart City', *First Monday*, http://journals.uic.edu/ojs/index.php/fm/article/view/5903/4660

Salame, Richard (2018) 'The New Taylorism', *Jacobin*, 2 February, www.jacobinmag.com/2018/02/amazon-wristband-surveillance-scientific-management

Sandifer, Elizabeth and Jack Graham (2018) *Neoreaction a Basilisk: Essays on and Around the Alt-Right*, CreateSpace Independent Publishing Platform.

Sanger, David (2018) *The Perfect Weapon: War, Sabotage, and Fear in the Cyber Age*, New York: Crown.

Satariano, Adam and Jack Nicas (2018) 'E.U. Fines Google $5.1 Billion in Android Antitrust Case', *New York Times*, 18 July, www.nytimes.com/2018/07/18/technology/google-eu-android-fine.html

Scharre, Paul (2018) *Army of None: Autonomous Weapons and the Future of War*, New York: Norton.

Schiller, Dan (1999) *Digital Capitalism: Networking the Global Market System*, Cambridge MA: MIT Press.

Schwab, Klaus (2017) *The Fourth Industrial Revolution*, Geneva: World Economic Forum.

Scola, Nancy (2013) 'Obama, the "Big Data" President', *Washington Post*, 14 June, https://preview.tinyurl.com/yd9kecaj

Searle, John R. (1980) 'Minds, Brains, and Programs', *Behavioural and Brain Sciences* 3:3.

Sedlak, Mackenzie (2016) 'A Quick Deep Learning Tutorial', *IBM Blog*, www.ibm.com/blogs/watson/2016/10/quick-deep-learning-tutorial

Seligman, Lara (2018) 'Pentagon's AI Surge on Track, Despite Google Protest', *Foreign Policy*, 29 June, https://foreignpolicy.com/2018/06/29/google-protest-wont-stop-pentagons-a-i-revolution

Seth, Manav (2018) '2019: The Year of Chatbot Evolution', *Chatteron.io*, https://blog.chatteron.io/2019-the-year-of-chatbot-evolution-e73dc63d38f0

Shanahan, Murray (2015) *The Technological Singularity*, Cambridge MA: MIT Press.

Shanin, Sergii (2018) 'Is Computer Engineering Really Going to Be Automated?', *Hacker Noon*, 29 May, https://hackernoon.com/is-computer-engineering-really-going-to-be-automated-e6111412432e

Shaviro, Steven (2009) 'The Singularity is Here', in Mark Bould and China Miéville (eds), *Red Planets: Marxism and Science Fiction*, London: Pluto.

Shead, Sam (2017) 'Tech Giants are Fighting to Hire the Best AI Talent at the NIPS Conference in LA This Week', *Business Insider*, 4 December, www.

businessinsider.com/tech-giants-are-fighting-to-hire-the-best-ai-talent-at-the-nips-conference-this-week-2017-12

Shead, Sam (2018) 'Britain, France and Germany Fight It Out to Be Europe's AI Leader', *Forbes*, 26 April, www.forbes.com/sites/samshead/2018/04/26/britain-france-and-germany-fight-it-out-to-be-europes-ai-leader/#7a1a94a0232f

Shewan, Dan (2017) 'Robots Will Destroy Our Jobs – And We're Not Ready for It', *Guardian*, 11 January, www.theguardian.com/technology/2017/jan/11/robots-jobs-employees-artificial-intelligence

Silicoff, Sean (2018) 'She Looks Like a Human: Can She Be Taught to Think Like One Too?', *The Globe and Mail*, 7 September.

Silver, Beverly (2003) *Forces of Labor: Workers' Movements and Globalization since 1870*, Cambridge: Cambridge University Press.

Silver, David et al. (2016) 'Mastering the Game of Go With Deep Neural Networks and Tree Search', *Nature* 529.

Simonite, Tom (2015) 'Facebook Joins Stampede of Tech Giants Giving Away Artificial Intelligence Technology', *MIT Technology Review*, www.technologyreview.com/s/544236/facebook-joins-stampede-of-tech-giants-giving-away-artificial-intelligence-technology

Simonite, Tom (2017) 'What is Ray Kurzweil Up to at Google? Writing Your Emails', *Wired*, 2 August, www.wired.com/story/what-is-ray-kurzweil-up-to-at-google-writing-your-emails

Simpson, Meagan (2018) 'IBM's New Bias-detection Software, Democratizing AI', *IT World Canada*, 24 September, www.itworldcanada.com/article/how-ibms-new-bias-detection-software-has-already-helped-companies/409206

SIX (n.d.) 'The Internet of "Bring!": Where Does the Fridge Order its Milk?', www.six-group.com/en/home/blog/internet-of-bring.html

Skidelsky, Robert (2018) 'Will the Population Become Redundant?', *Eurozine*, 9 November, www.eurozine.com/will-population-become-redundant

Smith, Jason E. (2017) 'Nowhere to Go: Automation, Then and Now: Part 1', *The Brooklyn Rail*, 1 March, https://brooklynrail.org/2017/03/field-notes/Nowhere-to-Go

Söderberg, Johan (2008) *Hacking Capitalism: The Free and Open Source Software Movement*, New York: Routledge.

Solon, Olivia (2018) 'Google's Robot Assistant Now Makes Eerily Lifelike Phone Calls For You', *Guardian*, 8 May, www.theguardian.com/technology/2018/may/08/google-duplex-assistant-phone-calls-robot-human

Solow, Robert (1987) 'We'd Better Watch Out', *New York Times Book Review*, 12 July, www.standupeconomist.com/pdf/misc/solow-computer-productivity.pdf

Srnicek, Nick (2016) *Platform Capitalism*, New York: John Wiley.

Srnicek, Nick (2018) 'Platform Monopolies and the Political Economy of AI', in John McDonnell (ed.), *Economics for the Many*, London: Verso.

Srnicek, Nick and Alex Williams (2015) *Inventing the Future: Postcapitalism and a World Without Work*, London: Verso.

Srivastava, Tavish (2015) 'Difference Between Machine Learning and Statistical Modeling', *Analytics Vidhya*, www.analyticsvidhya.com/blog/2015/07/difference-machine-learning-statistical-modeling

Standing, Guy (2011) *The Precariat: The New Dangerous Class*, London: Bloomsbury.

Statista (2016) 'Artificial Intelligence Market Revenue Worldwide 2015–2024', *Artificial Intelligence (AI) Statista Dossier*, www.statista.com/statistics/621035/worldwide-artificial-intelligence-market-revenue

Steinhoff, James (2014) 'Transhumanism and Marxism: Philosophical Connections', *Journal of Evolution and Technology* 24:2.

Steinhoff, James (2017) 'Artificial Intelligence and the Reconfiguration Thesis', paper presented at 'Marxism and Revolution Now', Marxist Literary Group Summer Institute on Culture and Society, University of California Davis, 24–28 June.

Steinhoff, James (2019a) 'Reservoirs of Cognition: Capital's Theory of AI as Utility', *Digital Culture and Society*.

Steinhoff, James (2019b) 'Labour Process Analysis and the Automation of Machine Learning Labour', paper presented at DIGISUM Critical Digital and Social Media Conference, 6–8 March, Umeå University, Umeå, Sweden.

Stiegler, Bernard (2010) *For a New Critique of Political Economy*, Cambridge: Polity.

Stiegler, Bernard (2015) *States of Shock: Stupidity and Knowledge in the 21st Century*, Cambridge: Polity.

Stiegler, Bernard (2017) *Automatic Society Volume 1: The Future of Work*, Cambridge: Polity.

Stone, Zak (2016) 'Celebrating Tensorflow's First Year', *Google Research Blog*, https://research.googleblog.com/2016/11/celebrating-tensorflows-first-year.html

Streitfeld, David (2018) 'Protesters Block Google Buses in San Francisco, Citing "Techsploitation"', *New York Times*, 31 May, www.nytimes.com/2018/05/31/us/google-bus-protest.html

Streitfeld, David and Mallia Wollan (2014) 'Tech Rides are Focus of Hostility in Bay Area', *New York Times*, 31 January, www.nytimes.com/2014/02/01/technology/tech-rides-are-focus-of-hostility-in-bay-area.html?_r=0

Strickland, Eliza (2018) 'Layoffs at Watson Health Reveal IBM's Problem With AI', *IEE Spectrum*, 25 June, https://spectrum.ieee.org/the-human-os/robotics/artificial-intelligence/layoffs-at-watson-health-reveal-ibms-problem-with-ai

Stross, Charles (2005) *Accelerando*, New York: Ace Books.

Sun, Ron (2014) 'Connectionism and Neural Networks', in Keith Frankish and William M. Ramsey (eds), *The Cambridge Handbook of Artificial Intelligence*, Cambridge: Cambridge University Press.

Sun, Yitang (2017) 'In China, a Store of the Future – No Checkout, No Staff', *MIT Technology Review*, 16 June, www.technologyreview.com/s/608104/in-china-a-store-of-the-future-no-checkout-no-staff

Sutton, Richard S. and Andrew G. Barto (1998) *Reinforcement Learning: An Introduction*, Cambridge MA: MIT Press.

Suvin, Darko (1979) *Metamorphoses of Science Fiction: On the Poetics and Literary History of a Literary* Genre, New Haven: Yale University Press.

Swanson, Brett and Micael Mandel (2017) 'Robots Will Save the Economy', *The Wall Street Journal*. May 14, www.wsj.com/articles/robots-will-save-the-economy-1494796013

Tam, Donna (2014) 'Meet Amazon's Busiest Employee – the Kiva Robot', *Cnet* Nov 30, www.cnet.com/news/meet-amazons-busiest-employee-the-kiva-robot

Tawil-Souri, Helga (2012) 'It's Still About the Power of Place', *Middle East Journal of Culture and Communication* 5.

Taylor, Astra (2018) 'The Automation Charade', *Logic*, https://logicmag.io/05-the-automation-charade

Tech Workers Coalition (2018) 'Tech Workers, Platform Workers, and Workers' Inquiry', *Notes From Below*, 30 March, https://notesfrombelow.org/article/tech-workers-platform-workers-and-workers-inquiry

Tegmark, Max (2017) *Life 3.0: Being Human in the Age of Artificial Intelligence*, New York: Knopf.

Tenenbaum, Joshua B., Charles Kemp, Thomas L. Griffiths and Noah D. Goodman (2011) 'How to Grow a Mind: Statistics, Structure, and Abstraction', *Science* 331:6022.

Terekhova, Maria (2017) 'JP Morgan Takes AI Use to the Next Level', *Business Insider*, 2 August, www.businessinsider.com/jpmorgan-takes-ai-use-to-the-next-level-2017–8

Thibodeau, Patrick (2016) 'One in Three Developers Fear A.I. Will Replace Them', *Computerworld*, 8 March, www.computerworld.com/article/3041430/it-careers/one-in-three-developers-fear-ai-will-replace-them.html

Thompson, James (2018) 'Why Artificial Intelligence is "Like Electricity" for Microsoft', *Financial Review*, www.afr.com/technology/why-artificial-intelligence-is-like-electricity-for-microsoft-20181102-h17fei

Tiqqun (2001) 'The Cybernetic Hypothesis', *Tiqqun* 2, http://cybernet.jottit.com

Torres, Phil (2018) 'Superintelligence and the Future of Governance: On Prioritizing the Control Problem at the End of History', in Roman V. Yampolskiy (ed.), *Artificial Intelligence Safety and Security*, New York: Chapman and Hall/CRC.

Toscano, Alberto (2011) 'Logistics and Opposition', *Mute* 3:2, www.metamute.org/editorial/articles/logistics-and-opposition

Toscano, Alberto (2014) 'Lineaments of the Logistical State', *Viewpoint Magazine*, www.viewpointmag.com/2014/09/28/lineaments-of-the-logistical-state

Tozzi, Christopher (2017) *Fun and Profit: A History of the Free and Open Source Software Revolution*, Cambridge MA: MIT Press.

Tractica (2018) 'Artificial Intelligence Edge Device Shipments to Reach 2.6 Billion Units Annually by 2025', www.tractica.com/newsroom/press-releases/artificial-intelligence-edge-device-shipments-to-reach-2–6-billion-units-annually-by-2025

Trichter, Judd (2015) *Love and Death in the Age of Mechanical Reproduction*, New York: Thomas Dunne.

Tronti, Mario (1977) *Ouvriers et Capital*, Paris: Christian Bourgeois.

US Department of Labour (2015) 'Occupational Employment Projections to 2024', Bureau of Labor Statistics, Monthly Labour Review, December, www.

bls.gov/opub/mlr/2015/article/occupational-employment-projections-to-2024.htm

US Department of Labor (2018) *Occupational Outlook Handbook, Software Developers*, www.bls.gov/ooh/computer-and-information-technology/software-developers.htm

US Equal Employment Opportunity Commission (USEEOC) (2016) 'Diversity in High Tech', www.eeoc.gov/eeoc/statistics/reports/hightech

Vast Abrupt (2018) 'Ideology, Intelligence, and Capital: An Interview With Nick Land', https://vastabrupt.com/2018/08/15/ideology-intelligence-and-capital-nick-land

Vercellone, Carlo (ed.) (2006) *Capitalismo cognitivo*, Roma: Manifestolibri.

Vincent, James (2017) 'Robots and AI Are Going to Make Social Inequality Even Worse, Says New Report', *The Verge*, 13 July, www.theverge.com/2017/7/13/15963710/robots-ai-inequality-social-mobility-study

Vincent, James (2018) 'Welcome to the Automated Warehouse of the Future', *The Verge*, 26 May, www.theverge.com/2018/5/8/17331250/automated-warehouses-jobs-ocado-andover-amazon

Virilio, Paul (2000) *The Information Bomb*, London: Verso.

Virno, Paolo (2004) *A Grammar of the Multitude*, Los Angeles: Semiotext(e).

Vitti, Joseph J. (2013) 'Cephalopod Cognition in an Evolutionary Context: Implications for Ethology', *Biosemiotics* 6:3.

Viv (n.d.) viv.ai, http://viv.ai

Vorhies, William (2016a) 'Data Scientists Automated and Unemployed by 2025!', *Data Science Central*, 13 April, www.datasciencecentral.com/profiles/blogs/data-scientists-automated-and-unemployed-by-2025

Vorhies, William (2016b) 'Deep Learning for Everyone – And (Almost) Free', *Data Science Central*, 18 August, www.datasciencecentral.com/profiles/blogs/deep-learning-for-everyone-and-almost-free

Vorhies, William (2017) 'Data Scientists Automated and Unemployed by 2025 – Update!', *Data Science Central*, 17 July, www.datasciencecentral.com/profiles/blogs/data-scientists-automated-and-unemployed-by-2025-update

Vr, Sridharan (2016) 'Bots for Messenger: Adding Analytics and FbStart Opportunities', *Facebook Developer News*, 14 November, https://developers.facebook.com/blog/post/2016/11/14/bots-for-messenger-adding-analytics-and-fbstart-program

Walker, Jon (2018) 'Machine Learning in Manufacturing – Present and Future Use-Cases', *TechEmergence*, www.techemergence.com/machine-learning-in-manufacturing

Wang, Brian (2018) 'World Will See Less Robot Job Loss Through 2030 as Robot Automation Leaders Foxconn and Tesla Face Delays', *The Next Big Future*, 24 April, www.nextbigfuture.com/2018/04/world-will-see-less-robot-job-loss-through-2030-as-robot-automation-leaders-foxconn-and-tesla-face-delays.html

Wang, Pei and Ben Goertzel (2007) 'Introduction: Aspects of Artificial General Intelligence', *Proceedings of the 2007 conference on Advances in Artificial*

General Intelligence: Concepts, Architectures and Algorithms: Proceedings of the AGI Workshop 2006.

Wark, McKenzie (2017) 'On Nick Land', www.versobooks.com/blogs/3284-on-nick-land

Watts, Peter (2018) *The Freeze-Frame Revolution*, San Francisco: Tachyon Publications.

Weaver, John Frank (2017) 'What Exactly Does It Mean to Give a Robot Citizenship?', *Slate*, https://slate.com/technology/2017/11/what-rights-does-a-robot-get-with-citizenship.html

Weber, Steve (2004) *The Success of Open Source*, Cambridge: Harvard University Press.

Wilde, Lawrence (2000) '"The Creatures, Too, Must Become Free": Marx and the Animal/Human Distinction', *Capital & Class* 72.

Williams, Alex and Nick Srnicek (2013) '#ACCELERATE MANIFESTO for an Accelerationist Politics', *Critical Legal Thinking Blog*, 14 May, http://criticallegalthinking.com/2013/05/14/accelerate-manifesto-for-an-accelerationist-politics

Wilson, Daniel H. (2012) *Robopocalypse*, New York: Vintage.

Wilson, Daniel H. (2015) *Robogenesis*, New York: Vintage.

Wilson, H. James and Paul R. Daugherty (2018) 'Why Even AI-Powered Factories Will Have Jobs for Humans', *Harvard Business Review*, 8 August, https://hbr.org/2018/08/why-even-ai-powered-factories-will-have-jobs-for-humans

Winfield, Alan (2012) *Robotics: A Very Short Introduction*, Cambridge MA: MIT Press.

Wingfield, Nick (2017) 'As Amazon Pushes Forward With Robots, Workers Find New Roles', *New York Times*, 10 September, www.nytimes.com/2017/09/10/technology/amazon-robots-workers.html

Wisskirchen, Gerlind et al. (2017) 'Artificial Intelligence and Robotics and their Impact on the Workplace', IBA Global Employment Institute, www.ibanet.org/Document/Default.aspx?DocumentUid=c06aa1a3-d355

Woland, Blaumachen and Friends (2014) 'From Sweden to Turkey: The Uneven Dynamics of the Era of Riots', *SIC International Journal of Communisation* 2.

Wong, Julia Carrie (2017a) 'Tesla Factory Workers Reveal Pain, Injury and Stress: "Everything Feels Like the Future But Us"', *Guardian*, 18 May, www.theguardian.com/technology/2017/may/18/tesla-workers-factory-conditions-elon-musk

Wong, Julia Carrie (2017b) 'Tesla Workers Were Seriously Hurt More than Twice as Often as Industry Average', *Guardian*, 24 May, www.theguardian.com/technology/2017/may/24/tesla-factory-workers-injuries-higher-than-industry-average

Wood, Zoe (2018) 'Rise of Robots Threatens to Terminate the UK Call-centre Workforce', *Guardian*, 12 May, www.theguardian.com/business/2018/may/12/robot-technology-threat-terminist-uk-call-centre-workforce

Woodcock, James (2017) *Working the Phones: Control and Resistance in Call Centres*, London: Pluto.

World Economic Forum (WEF) (2016) 'The Future of Jobs Employment, Skills and Workforce Strategy for the Fourth Industrial Revolution', http://www3.weforum.org/docs/WEF_Future_of_Jobs.pdf

World Economic Forum (WEF) (2017) 'Impact of the Fourth Industrial Revolution on Supply Chains', White Paper, Geneva, Switzerland.

Wykstra, Stephanie (2018) 'Just How Transparent Can a Criminal Justice Algorithm Be?', *Slate*, July 3, https://slate.com/technology/2018/07/pennsylvania-commission-on-sentencing-is-trying-to-make-its-algorithm-transparent.html

Xiang, Feng (2018) 'AI Will Spell the End of Capitalism', *Washington Post*, 3 May, www.washingtonpost.com/news/theworldpost/wp/2018/05/03/end-of-capitalism/?utm_term=.17d7c9d56637

Yampolskiy, Roman V. (n.d.) 'Artificial Intelligence Safety and Cybersecurity: A Timeline of AI Failures', *arxiv.org*, https://arxiv.org/ftp/arxiv/papers/1610/1610.07997.pdf

Ying, Wang (2018) 'Industrial Robot Sales in China Hit Record', *Telegraph*, 13 July, www.telegraph.co.uk/news/world/china-watch/technology/industrial-robots

Yeginsu, Ceylan (2018) 'If Workers Slack Off, the Wristband Will Know (And Amazon Has a Patent for It)', *New York Times*, 1 February, www.nytimes.com/2018/02/01/technology/amazon-wristband-tracking-privacy.html

Yu, Miaomiao (2017) 'The Humans Behind Artificial Intelligence', *Synced*, 30 April, https://syncedreview.com/2017/04/30/the-humans-behind-artificial-intelligence

Zamponi, Lorenzo (2018) 'Bargaining With the Algorithm', *Jacobin*, 9 June, www.jacobinmag.com/2018/06/deliveroo-riders-strike-italy-labor-organizing

Zeifman, Igal (2017) 'Bot Traffic Report 2016', *Incapsula*, www.incapsula.com/blog/bot-traffic-report-2016.html

Zerowork Collective (1975) 'Introduction', *Zerowork: Political Materials* 1.

Zhang, Yu-Dong et al. (2017) 'Facial Emotion Recognition Based on Biorthogonal Wavelet Entropy, Fuzzy Support Vector Machine, and Stratified Cross Validation', *IEEE Access* 4.

Zilis, Shivon and James Cham (2016) 'The Current State of Machine Intelligence 3.0', *O'Reilly*, www.oreilly.com/ideas/the-current-state-of-machine-intelligence-3–0

Index

1844 Manuscripts 115–17
2312 (Robinson) 26

abstract labour 136
abstraction 66
abyssal view 8
Accelerando (Stross) 27, 137
accelerationist view 7 *see also* maximalist view
Accenture 59, 73
accumulation 31, 43, 44, 141, 143
actually-existing AI 10, 11–12
actually-existing AI-capitalism 2, 21, 50, 51tab, 146
Adami, Christoph 126
adaptability 118, 119, 124–5, 126
advertising 33, 37, 98–9, 100
affect 64, 66
agency 139
AGI (artificial general intelligence)
  concept of 10, 110, 111–13
  Marxist theory and 130–8
  narrow AI approaches to 127–9
  Neoreaction and 157–8
  personification of capital and 138–9
  projects on 1, 113–14
  proletarianization of 135–8
  surplus population and 140–4
AI (artificial intelligence)
  actually-existing AI 10, 11–12
  communist orientation to 146–56, 160–2, 166n1
  definitions and types 9–15
  'democratization' of 52–6
  goal of 3, 9, 32
  key players 32–8, 38–42
  as technical fix to capital's problems 71–2, 73, 74–5
  see also ML (machine learning)
'AI Apocalypse Now' view 87–8, 91, 142
AI Effect 9
AI Safety movement 44–5
AI science fiction 24–8
Alexa 34, 58, 140
Algorithmic Justice League 105
algorithms
  common current uses of 33, 36–8
  discrimination and 86–7, 95–6, 104, 105–6, 164n2
  for hiring workers 92, 95–6
  model building and deployment 76–9
  resistance to 104–6
  for workplace surveillance 92–4
Allstate insurance 83
Alpaydin, Ethem 114, 127
Alphabet *see* Google/Alphabet
AlphaGo 14, 37, 88, 120–1, 123
Amazon
  AI-based products, range of 37
  Alexa 34, 58, 140
  automated stores 59–60, 61–2, 85
  crowd sourcing platform of 78
  employee resistance at 101–2, 103
  fulfilment centres of 84–5, 93, 101–2
  Web Services (cloud) 42, 53
ambient intelligence 57–62
Andreessen, Marc 150
Android operating system 37, 54–5, 99
androids 26, 68, 131, 137
animals 115–20, 165n9
Apache 2.0 53
Apple 33, 36, 38, 74, 163n2
artificial general intelligence *see* AGI
artificial intelligence *see* AI

artificial neural networks (ANNs) 13, 166n13
convolutional 120
deep 76, 120
generative 123
generative adversarial network 65
recurrent 58
recursive cortical network 123
Ashton, Kevin 56–7
ASI (artificial superintelligence) 10–11
'augmentation' euphemism 83, 90
augmented reality (AR) 33
automata (historic) 130–1
automated luxury communism 146–7
AutoML 129–30
autonomist Marxism 69–75, 155–6
autonomous vehicles 33, 42–3, 45, 81, 82–3
Azure cloud service 42

Baidu 33, 35, 42, 53–4, 80
Banks, Ian 26
Bastani, Aaron 6
Baum, Seth D. 11, 113–14, 165n5, 165n6
BAYOU ML 79
Benanav, Aaron 141–2
Bennett, Jane 161
Benton, Ted 119
Bernes, Jasper 147–9, 153
Bezos, Jeff 1
Blade Runner 25–6
Boeing 159
Bost, Matthew W. 67
Bostrom, Nick 11, 129, 151
Bratton, Benjamin 112, 152
Brexit 101
Brin, Sergey 1, 4, 37
Brooks, Rodney 12
Brynjolfsson, Erik 87–8
bubbles 44–6
'Business-as-Usual' view 87, 88–9, 91, 109, 142
business investment, since 2008 crash 45–6, 80

Caffe2 53
Caffentzis, George 19, 23–4, 132
call centres 83–4, 93
Cambridge Analytica 100
Čapek, Karel 133–5
capital
AI as technical fix for problems of 71–2, 73, 74–5
circulating 70–4, 79–80, 82–7
constant 15–21
falling rate of profit and 17–18, 19
fixed 15–21, 16, 63–4, 69–70, 79–80, 133, 134, 137, 163n5, 164n5
freedom from 155–6
humans, emancipation from 7, 32, 111, 140–4, 145, 149, 157–9
organic composition of 17, 19, 24, 141, 155
personification of 138–9
state support of 38–42
variable 15, 17, 18, 69–70, 138–9
CAPTCHA codes 122–3, 124
caregiving 95
casino workers 68
chatbots 58–9, 60, 83, 92
chess (Deep Blue) 12, 36, 127
China
AI development by 40–1, 80
automated stores in 85
communist views of AI in 166n1
labour issues in 74, 81
resistance in 102
US competition with 35, 40–1
Chinese Room experiment 11, 126
choice (by humans), replacement of 56–8, 98, 140, 145
circulation of capital 70–4, 79–80, 82–7
citizens' income (universal basic income, UBI) 6, 143, 146, 150–1
citizenship 137
class composition 69–75, 79
class conflict 16–17, 101
class power 69, 73
Clegg, John 141–2

closed source creep 54–6
cloud (the) 42–3
CNN (convolutional neural networks) 120, 165n12
CNTK 53
Cockshott, Paul 119, 125
cognition
 ambient intelligence and 58–60
 ANNs and 13
 GOFAI and 11–12
 human, as a restraint 52, 57
 Marx on 117–18, 119
 means of 31, 62–7
 SED and 12
 use of term 31, 52
cognitive capitalism 31, 63 *see also* post-Fordism
Collins, Harry 24–5
commodities 132–4, 137–8
communication 58–60, 65 *see also* replacement: of human interaction
communism of capitalism 56
communist orientation to AI 149–56, 160–2, 166n1
competition 17
compulsion 135
concentration 41, 42, 43–4, 107–8
consciousness 11, 113, 116–17, 126–7, 139
constant vs fixed capital 15–21
consumption 17, 18 *see also* circulation of capital
content moderation 78
control function 23
convolutional neural networks (CNN) 120, 165n12
Cook, Mike 25
corvée 135
creativity 110, 117, 118, 120–1, 126
crises/recessions 17–18, 21, 39, 45–6, 73, 142
'Culture' novels (Banks) 26
curation 61–2
cybernetic capitalism 22–3, 50, 51tab
cyberpunk science fiction 25–6
dark-side AI 55–6
Dartmouth College workshop 9, 12, 36, 111
data cleaning and moderating 77–8
data protection 41, 104–5
De Stefano, Valerio 91–2
decomposition of working-class 70, 71–2, 73, 75, 97–101
Deep Blue 12, 36, 127
deep learning 66, 127, 164n2
deep neural networks 76
DeepMind Technologies 1, 14, 35, 37, 113, 114, 120–1, 123
defence industry 39–40, 74–5, 103, 152–3 *see also* security industry
Deliveroo 102
Department of Defense (US) 39, 40
Didi 102
digital capitalism *see* post-Fordism
digital journalism 102
digital personal assistants 33
discrimination 86–7, 95–6, 103, 104, 105–6, 157–8, 164n2
disposability of populations 5–6
dispossession 138
Domingos, Pedro 127
Dong, Catherine 76–7
dot.com bubble 44
Drexler, Eric K. 151
driverless vehicles 33, 42–3, 45, 81, 82–3
DSSTNE 53
Duplex 58–9, 83–4

economic crises 17–18, 21, 39, 45–6, 73, 142
The Economist 33, 34, 43–4
ecosocialism 161–2
edge computing 42–3, 99
education 96–7
elder care 95
Elster, Jon 119
emancipation, of capital from humans 7, 32, 111, 140–4, 145, 149, 157–9
emotions 66

employees of tech companies *see* tech industry
environment 152
Ernest, Guy 53
Eubanks, Virginia 86–7
European Union 41, 104–5, 137
existential risk 151–2
Expensify 77–8
expert systems 12, 32 *see also* GOFAI
exploitation 3, 6, 15, 20, 69–70, 84, 102, 136, 149, 152

Facebook
AI research at 37, 98–9
bots on 59
content moderation at 21, 78
opening of libraries by 53
resistance to 105, 107
Trump campaign and 100–1, 107
facial identification software 103, 104
Fall Revolution books (MacLeod) 27
falling rate of profit (FROP) 17–18, 19
far right/neo-fascism 73, 100–1, 157–8
fauxtomation 5
FBLearner Flow (FBL) 98–9
Federated Learning project 99
Federici, Sylvia 5
Fei Fei Li 53
fetishism 139, 149
'Fight for 15' movement 5
Figure Eight 77
financial crisis (2008) 21, 39, 45–6, 73
financialization 45–6, 72, 73
fixed capital
constant capital vs 15–21
definition 163n5, 164n5
general social knowledge and 63–4 (*see also* general intellect)
machines as 133, 134
slaves as 137
social function of 16, 79–80
variable capital vs 15, 17, 18, 69–70
flexibility 110, 118, 125
Ford, Martin 87–8
Fordism 50, 51tab, 70
forms of appearance 132–3 *see also* social forms
FOSS (free and open-source) programming 53–6, 113
Foster, John Bellamy 161
Foxconn 74, 81
'Fragment on Machines' (Marx) 18–19, 67, 109, 131–5, 136, 152–3
Frank, Morgan R. 91
freedom 116–18, 136–7
Frey, Carl 88
FROP (falling rate of profit) 17–18, 19

Gartner 44, 59, 83
Gates, Bill 4
GDP, labour's share of 73
GDPR (General Data Protection Regulation) 41, 104–5
gender
abolition of 167n4
composition of labour and 72
discrimination on 95–6, 103, 105–6, 157–8
simulation of 58
general conditions of production
AGI and 129–30
ambient intelligence and 58–62
concept of 30–1
open-source and 53–6
the state and 39–40
General Data Protection Regulation (GDPR) 41, 104–5
general intellect 31–2, 63–7
general intelligence 125–6, 128 *see also* AGI
generalization 48, 121
gig economy 93, 150–1
Gillespie, Tarleton 38
GitHub 79
global warming 152
globalization 71–4, 77–8, 81
Go (game) *see* AlphaGo
Goertzel, Ben 110, 112–13, 128–9
GOFAI (Good Ol' Fashioned AI) 11–12, 32

Golumbia, David 158
Google/Alphabet
automated machine learning project 129–30
chatbots by 58–9
cloud service of 42, 55
content moderation by 78
DeepMind 1, 14, 35, 37, 113, 114, 120–1, 123
'democratization' of AI and 53, 54–5
discriminatory hiring and 96
employee resistance at 40, 103
new sectors targeted by 33
online booking system of 83–4
overview of AI activities 36–7
public resistance to 106
search algorithm of 98
Singularity U 159
smart cities and 57, 106–7
smartphones, use of for AI training 99
state support of 39
Gordon, Robert 45
Gourevitch, Alex 151
Grace, Katja 121
grammatization 97–101
'grey goo' example 151
Grundrisse (Marx) 18–19, 46–9 *see also* 'Fragment on Machines'
Gubrud, Mark Avrum 111
Gui, Terry 74, 81

Haldane, Andy 89
HANA 34
Hanson Robotics 137
Harari, Yuval Noah 126–7, 151
hardware 42–3
Hester, Helen 58
HLAI (human-level artificial intelligence) 128
HLMI (human-level machine intelligence) 112, 114, 165n3, 165n4
hospitality workers 68
Hughes, Chris 150
Human Brain Project 113
Human Rights Data Analysis Group (HRDAG) 105
Humans (TV show) 133

IBM 12, 36, 39, 42, 53, 66, 127
ICE (US Immigration and Customs Enforcement agency) 103
ILO (International Labour Organization) study 91
imagination 117, 118, 119, 121–3
Impett, Leo 164n2
inductive biases 122–3
industrial automation 74, 81–2, 90–1
industrial capitalism 51tab *see also* Fordism; large-scale industry
Industry 4.0 81 *see also* Internet of Things
infrastructure
AI as 30–1, 52–6, 61–2
as general condition of production 30–1, 39, 46–9, 61–2
privatization of 39
integral accidents 151–2
intensification of work 17, 21
Internet of Things (IoT) 56–7, 62, 81
investment, since 2008 crash 45–6, 80
The Invisible Committee 147–8

Jameson, Frederic 149
Japan 22, 41, 95
Jenkins, Simon 89
Jeopardy 36
Jordan, Tim 78
journalism 102

Kai-Fu Lee 40–1, 109
Kaplan, Jerry 9, 11, 14–15
Kasparov, Garry 12, 36, 127
Kelly, Kevin 30, 61
Kiva Systems 84
Kjøsen, Atle Mikkola 58– 59, 131
knowledge capture 64
knowledge systems 12
Kurzweil, Raymond 1, 37, 129, 156, 159, 165n1

labour
Marx's concept of 110, 115–20, 124–7, 135–7
replacement of 58–60, 78–9, 82–94, 135–44
share of GDP 73
see also workers
labour theory of value 110, 132
Lake, Brenden M. 121
Land, Nick 7, 153, 156–8
Landing.ai 34
large-scale industry 49, 51tab
learning biases 122–3
Lebeuf, Carlene 59
left-accelerationist view 7, 146–56
legal status 136–7
Lenin, Vladimir 6
LinkedIn, discriminatory hiring and 96
Linux 54
lobbying 41
logistical capitalism *see* post-Fordism
logistics revolution 82–3
Lucas Plan 154
Luddism 166n2
luxury automated communism 146–7

machine learning (ML) *see* ML
MacLeod, Ken 27
Maluuba 113, 114
Mandel, Ernest 22
manufacture epoch 48–9, 51tab
manufacturing, automation of 74, 81–2, 90–1
Manzerolle, Vincent 58–9
Marin, Michael 53
Marks & Spencer 83
Marx, Karl
architect and bee example 117, 119, 123–4
on circulation of capital 70–1
on concentration 43
on crises 17–18, 21
on general conditions of production 30–1, 39, 46–9, 61, 129
on general intellect 31–2, 63–4, 125–6
on inhuman power 2, 4
on labour and labour-power 110, 115–20, 124–7, 135–7
on machinery 15–21, 129, 147, 152–3; on automation 18–21, 62, 67, 69–70, 129; value, social forms and 130–5, 166n16
on surplus population 141, 142, 144
Mason, Paul 6
Mason, Zachary 27
Maven (Project) 103
maximalist view ('it's really happening, let's speed it up') 6–7, 145, 146–56
McAfee, Andrew 87–8
McCorduck, Pamela 9
means of cognition 31, 62–7
Mechanical Turk (MTurk) 78, 102
Media Mobilizing Project 105–6
Melville, Andrew 60
mercantilism 51tab
Mercer, Rebekah 100
Mészáros, István 160
Microsoft
AGI Project 113, 114
algorithm checking by 78
automated retailing and 86
cloud service of 42
current activities (overview) 38
'democratization' of AI and 52–3
employee resistance at 103
vision of AI 30
middle class, hollowing out of 94–5
Mighty AI 77
military-industrial complexes 39–40, 74–5, 152–3
minimalist view 4–6, 145–6
ML (machine learning)
AGI and 128–30
definition and types 11, 12–15
Marx's visions and 21
oligopolies based on 32–8
production of 76–9
model building 76

monopolies *see* oligopolies/monopolies
Moody, Kim 80
Moore, Jason 161
Moore, Phoebe 92–3
Moravec, Hans 12, 159
Morris-Suzuki, Tessa 22, 23
Morton, Timothy 119, 161
MTurk (Mechanical Turk) 78, 102
Musk, Elon 1, 4, 81, 113, 159

Narrative Science 65
narrow AI 2, 10, 11–12, 88, 127–9
National Security Agency (US) 100
natural form 132–3 *see also* use-value
Natural Language Generation (NLG) 65
necessary labour-time 136
Negarestani, Reza 166n3
neo-fascism/far right 73, 100–1, 157–8
Neoreaction 157–8
Neuralink 159
news stories, automated 65
Ng, Andrew 30, 34, 127
Nicolaus, Martin 67
Nilsson, Nils J. 128, 165n4
novum 25

Obama, Barack 101
object recognition 45, 122–4
Ocado 85
OECD studies 90
oligopolies/monopolies 41, 42, 43–4, 107–8
Omohundro, Stephen M. 137–8
one-shot learning 123
O'Neill, Cathy 86
open-source programming 53–6, 113
OpenAI 1, 65–6, 113
operaismo 67, 156 *see also post-operaismo*
O'Reilly, Tim 54, 150
organic composition of capital 17, 19, 24, 141, 155
Osborne, Michael 88

Page, Larry 1, 37
Panzieri, Raniero 155–6
paper clip example 151–2
Pasquinelli, Matteo 164n7
people analytics 92–7
perception 58–60, 67
perpetuum mobile 130–1
Pichai, Sundar 103
Pitts, Frederick Harry 19
platform capitalism 36, 55–6, 93–4
Playment 77
Plum Rains (Romano-Lax) 26
policing 87, 105
political composition 70
politics 99–101, 107–9
post-Fordism 50, 51tab, 64, 71–2
post-operaismo 31, 63, 64, 67, 164n7
postcapitalism 6, 19, 26, 146–9, 150, 151, 154, 155
posthuman feminism 161, 167n4
power *see* class power
primitive accumulation 143
production, automation of 74, 81–2, 90–1
productivity 16, 45–6
profit
  falling rate of (FROP) 17–18, 19
  globalization and 71–4
  surplus-value and 24
proletarianization 22, 23, 135–8
PyTorch 53

Quill 65

racism and racial discrimination 87, 95–6, 104, 105–6, 157–8
radio-frequency identification (RFID) 56–7
Ramtin, Ramin 22–3
RCN (recursive cortical networks) 123–4
recessions/crises 17–18, 21, 39, 45–6, 73, 142
recommendations (algorithmic) 33, 37, 76, 84
reconfiguration debate 147–9

recursive cortical networks (RCN) 123–4
refusal, doctrine of 67, 155–6
regulation 40, 41, 55, 104–5
reification 149
reinforcement learning 14
Renaud, Karen 119, 125
resistance
by Amazon employees 101–2, 103
by Foxconn employees 74
by general population: to AI and surveillance 104–5, 147–8; to capital's power 73, 99–100; to tech companies 106–7
by Google employees 40, 103
'Results of the Immediate Process of Production' (Marx) 19–21
retail
automated purchasing 140
automated stores 59–60, 61–2, 85–6
warehousing 84–5, 93, 101–2
revolution 17, 18, 153–5
Rich, Elaine 9
right-wing extremism 73, 100–1, 157–8
Roberts, Michael 46
Roberts, Sarah 78
Robinson, Kim Stanley 26
robots
AI vs 2, 9–10
aspirations for 1
in industrial production 74, 81–2, 91
SED and 12
in service sector 68, 95
in warehouses 84–5
Romano-Lax, Andromeda 26
Rose, Geordie 1, 2
Rosenblat, Alex 94
R.U.R. (play) 133–5
Russia 41, 100

Salesforce 103
Samsung 1, 54–5
San Francisco, inequality in 106–7
Sanctuary Cognitive Systems Corporation 1
SAP 33–4
Saudi Arabia 137
Schmidt, Eric 107
science fiction 24–8, 68, 133–5
Searle, John 11, 126
security industry 80 *see also* defence industry
SED (situated, embodied and dynamical framework) 11, 12
Sedol, Lee 14, 88, 120–1
self-driving vehicles 33, 42–3, 45, 81, 82–3
sentiment analysis 66
serfdom 135
service sector jobs 68–9, 72, 82–7, 94–5, 101–2
Shanahan, Murray 127
shape recognition 122
Shewan, Dan 96
Sidewalk Labs 106–7
Siemens 81
Silicon Valley 106–7 *see also* tech industry
Silicon Valley De-Bug 105
Silver, Beverly 71
Singularity University 159
SIX 62
slavery 135, 137
Slow Tsunami view 143–4
smart cities 57, 106–7
smartphones 54–5, 99
Smith, Jason 142
Snowden, Edward 100
social brain 64
social factory 70–4
social forms 132–5
social individual 63–4
social media
defection from 105
subsumption of general behaviour and 98–101
see also specific media
social relations of production 132–3, 139

socialism 2, 146–9
socially necessary labour 17–18, 24
software developers *see* employees of tech industry
Solow, Robert 45
spatial fixes 71–2
speculation 44
Srnicek, Nick 6, 36, 148, 163n2
Stallman, Richard 54
Stanczyk, Lucas 151
state, the
  conflict with tech industry 164n4
  general conditions of production and 49
  policing, use of AI in 87, 105
  support of tech industry by 38–42, 74–5
  surveillance by 100, 107
  workers, moves against 72
Stiegler, Bernard 97–8
strikes 101–2
strong AI 11
Stross, Charles 26–7, 137
subsumption
  of general human behaviour 97–101
  hypersubsumption 21, 51table
  Marx on 19–21
  see also grammatization
Sunflowers (Watts) 27–8
supervised learning 14
supply-chain capitalism *see* post-Fordism
surplus labour-time 16
surplus populations 5, 23, 140–4
surplus-value
  expansion of as capital's goal 3, 31, 135–6
  fixed capital and 16, 79–80, 134
  profit and 24
  relative 20
surveillance
  military 40, 74–5, 80
  resistance to 104–5
  by the state 104–5, 147–8
  of workers 92–4
Suvin, Darko 25
Sweden, automated stores in 85
symbolic AI 11

taxi drivers 102
Taylor, Astra 4–6
Taylorism 50, 51tab, 70, 92
Teamsters Union 82
tech industry
  backlash against 107–8
  concentration in 41, 42, 43–4, 107–8
  contract/precarious workers in 77–8, 79, 102
  control of AI by 3–4
  culture of 157–8
  elite employees of: AI replacement of 78–9; competition for 34–5, 54, 55, 76; culture of 157–8; discrimination and 78–9, 95–6, 103; resistance by 40, 103; salaries for 35, 94, 103
  open-source, commodification of by 53–6
  state support of 39–42
  strengthening of by financial crisis 73–4
  UBI as reinforcement for 150–1
Tech Workers Coalition 103
technical composition of production 69–70
technical fixes 71–2, 73, 74–5
TechReset Canada 107
Techsploitation is Toxic 106
TensorFlow 53, 54–5
Tesla vehicles 81
Thiel, Peter 157
Thompson, James 30
Toronto 106–7, Google/Alphabet
Toscano, Alberto 147–9
Tractica 42
traffic *(verkehrs)* 67
transhumanism 159–61, 167n4
Trichter, Judd 26
trucking 82–3
Trudeau, Justin 107
Trump, Donald 40, 100–1
Turker Nation 102

Uber 82, 93, 102, 113, 114
unemployment 23–4, 87–91, 108–9, 140–4, 145 *see also* workers: replacement of
unions and worker coalitions 68, 82, 102, 103
universal basic income (UBI) 6, 143, 146, 150–1
unsupervised learning 14
urban development 57
use-value 71, 132, 134, 135–6 *see also* natural form
utilities *see* infrastructure

value 132, 134, 135–6
  socially necessary labour and 17–18, 24
  valorization of 139
  see also surplus-value
variable capital
  AGI as 138–9
  fixed capital vs 15, 17, 18, 69–70
  see also labour
verkehrs (traffic) 67
Vicarious PFC 122–4
Virno, Paolo 56, 64–5, 66–7
Viv (n.d.) 30
Void Star (Mason) 27

wages, suppression of
  during 1970s/80s 72
  by AI deployment 28–9, 80, 81, 142–3
  by economic crises 18, 21, 80
  as main issue today 93
  Marx on 17, 135–6
  mass movements to fight 70
  neo-fascism as result of 73
  resistance to 101–2
  surplus population and 141–2
  within tech industry 77, 93
  by threat of AI 5, 74
  women, minorities and 96
Walmart 34, 85–6
Wang, Pei 110, 128–9
Warp-CTC 53
Washington consensus 39
Watson (IBM) 36
Watts, Peter 27–8
weak AI 11
Westworld 68
white supremacists 157–8
Wilde, Lawrence 124
Williams, Alex 6, 148
Winfield, Alan 9–10
women *see* gender
workers
  algorithmic surveillance of 92–4
  class composition and 69–75
  doubly free 136–7
  intensification of labour for 83, 84, 85, 90
  Marx on machinery and 15–21
  polarization among 94–5
  precarity for 77–8, 93–4, 142–3, 150–1
  qualitative employment issues for 91–7
  replacement of 58–60, 78–9, 82–94, 135–44
  resistance by 40, 101–2, 103
  subsumption of 19–21
  in tech industry (*see* tech industry)
  threat of technology as weapon against 4–5, 17, 74
  training of 96–7
  see also wages, suppression of

xenofeminism 167n4

Y Combinator 150
Yampolskiy, Roman V. 44–5
Yang, Dan 150
Yudkowsky, Eliezer 11

Zuckerberg, Mark 1